RADIO SERVICING

VOLUME 3

F.M. Receivers and Audio Equipment

RADIO SERVICING

VOLUME 3

F.M. Receivers and Audio Equipment

G. N. PATCHETT
B.Sc.(Eng.), Ph.D., C.Eng., F.I.E.E., F.I.E.R.E., M.I.E.E.E.

LONDON
NORMAN PRICE (PUBLISHERS) LTD

NORMAN PRICE (PUBLISHERS) LTD
17 TOTTENHAM COURT ROAD, LONDON W1P 9DP

ISBN 0 85380 053 7

Printed in Great Britain by
Biddles Ltd., Guildford, Surrey.

AUTHOR'S PREFACE

THE Radio Servicing series was originally written for the City and Guilds of London Institute RADIO AND TELEVISION SERVICING course. This has been superseded by the MECHANICS' COURSE IN RADIO, TELEVISION AND ELECTRONICS 222, and the corresponding Technicians' course. (A series of books has been specially prepared for the Mechanics' Course 222 by the same author).

This volume is completely new and replaces the original Volume 3 (*Final Radio Theory*) by B. Fozard. It covers f.m. receivers and other electronic consumer devices (excluding television) *e.g.* record and tape equipment and hi-fi amplifiers.

The series is continually being brought up to date to keep pace with the rapid advances that take place in radio and allied equipment. The general title "Radio Servicing" is retained but the series in fact deals in detail with the theory of the radio receiver, etc. and considerable emphasis is placed on explaining in simple terms how the equipment operates. This is not always easy with modern sophisticated equipment but the author feels that to tackle difficult faults one must have a knowledge of how the equipment operates

CONTENTS

CHAPTER I

FREQUENCY MODULATION

THE amplitude modulation (a.m.) receiver was considered in *Radio Servicing*, Volume 2. In this volume the receiver designed for frequency modulation will be considered. Some receivers are, of course, designed for the reception of both a.m. and f.m. and these will be described. Although the basic principles of the receiver are not changed by the use of frequency modulation many changes are necessary. We still require r.f. amplification (which in a superhet receiver is obtained at both r.f. and i.f.); a demodulator; and a.f. amplification. We will first discuss what is meant by frequency modulation.

FREQUENCY MODULATION

In amplitude modulation the amplitude of the carrier is varied according to the modulation; in frequency modulation the frequency is varied according to the modulation, and the carrier amplitude is maintained constant. The principle of frequency modulation is shown in figure 1.1. At point A the modulating voltage is zero, hence the frequency of the carrier is at its normal

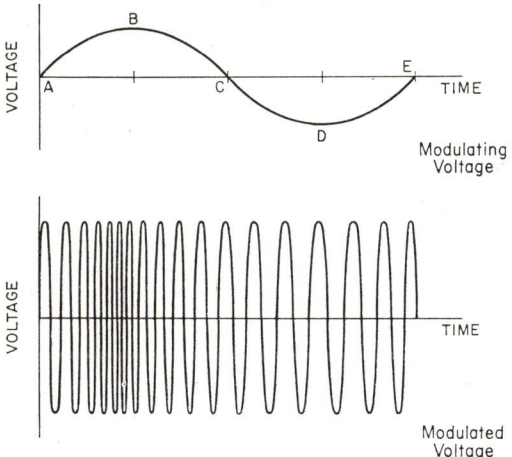

FIG. 1.1.   FREQUENCY MODULATION

value. The modulating voltage increases from A to a maximum, in a positive direction, at B. The carrier frequency is increased in proportion to the amplitude of the modulating voltage, and the frequency is therefore a maximum at point B. From B to C the modulating voltage is reduced to zero at C, and the frequency of the carrier decreases to its normal value at C. From C to D the modulating voltage increases in a negative direction, the carrier frequency decreasing to a minimum value at D. From D to E the modulating voltage

1

decreases to zero at E; the carrier frequency increases and returns to its normal value at E.

There are two misleading points as regards figure 1.1:

(a) The ratio of carrier frequency to modulating frequency is MUCH higher than that shown. For example, the modulating frequency may be 10,000 Hz and the carrier frequency 90 MHz (90,000,000 Hz). This means that there should be 9,000 cycles shown between A and E instead of some 18 cycles.

(b) The frequency variation is MUCH smaller than that shown. For example, the frequency variation may be 75 kHz for a carrier frequency of 90 MHz, *i.e.* the frequency variation is

$$\frac{75}{90,000} \times 100 \% = 0{\cdot}083\%.$$

Obviously, the diagram cannot be drawn to show this small frequency variation.

We will next see what the effect is of varying the amplitude and frequency of the modulating voltage. If the amplitude of the carrier is increased, as shown in figure 1.2, then the greater will be the carrier variation in frequency or what is called the **deviation of the carrier**. This is because the change in frequency is proportional to the amplitude of the modulating voltage. If the magnitude of the modulating voltage is maintained the same as in figure 1.2, but the frequency is doubled, the effect is as shown in figure 1.3. In this case the magnitude of the change in frequency is the same but the rate at which it is varied is doubled, *i.e.* it changes between its two limits in half the time. Thus the AMOUNT of carrier frequency change is proportional to the MAGNITUDE of the modulating voltage, but the RATE AT WHICH THE FREQUENCY is changed depends on the FREQUENCY of the modulating voltage.

As in amplitude modulation, frequency modulation results in the produc-

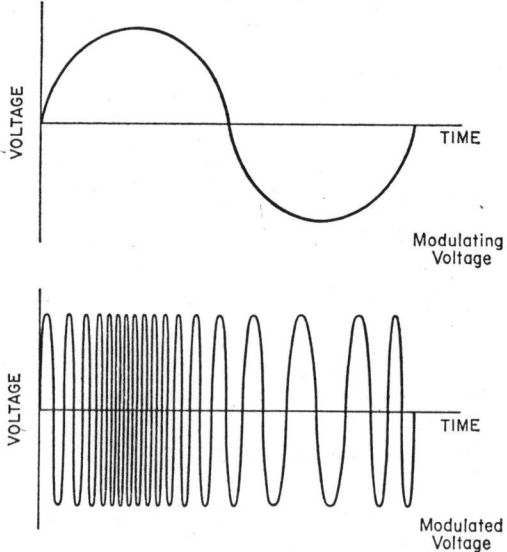

FIG. 1.2.   EFFECT OF LARGER AMPLITUDE OF MODULATING VOLTAGE THAN THAT SHOWN IN FIGURE 1.1.

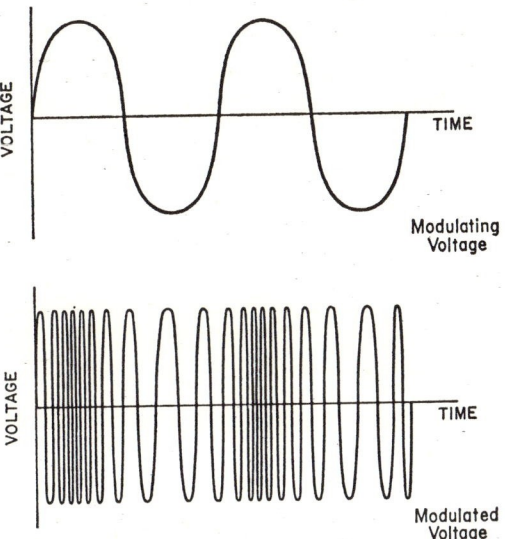

FIG. 1.3. EFFECT OF HIGHER FREQUENCY OF MODULATING VOLTAGE THAN THAT SHOWN IN FIGURE 1.2.

tion of sidebands. The arrangement of the sidebands is much more complicated. When the carrier is frequency modulated with a single frequency a number of sidebands are produced, not just two as in amplitude modulation. These sidebands are situated at multiples of the modulating frequency (f_m) from the carrier (f_c) and one example is shown in figure 1.4. In this case three sidebands have been shown each side of the carrier. Theoretically the side-

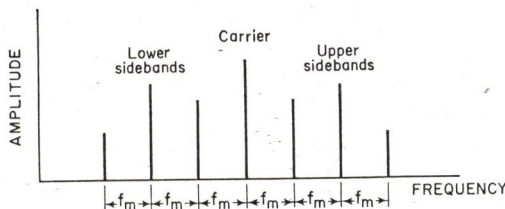

FIG. 1.4. TYPICAL CARRIER AND SIDEBANDS OF A FREQUENCY MODULATED SIGNAL

bands extend to infinity in both directions, but fortunately they rapidly decrease after a certain number: those below 1% are neglected. The way in which the sidebands are distributed and their amplitudes is complex and mathematically far beyond this book.

The number of sidebands depends on what is called the **modulation index,** and this is defined as

$$\frac{\text{Variation of carrier frequency}}{\text{Modulation frequency}}$$

The variation of carrier frequency will depend on the magnitude of the modulating voltage and on the maximum deviation of the system. In the v.h.f. radio service in this country the maximum deviation is $\pm75,000$ Hz (75 kHz). The modulation index is normally worked out for the maximum deviation as this results in the maximum number of sidebands and required bandwidth. If we consider a modulating frequency of 15,000 Hz (15 kHz), which is about the maximum transmitted, then the modulation index is:

$$\frac{75,000}{15,000} = 5$$

The calculation of the number of sidebands is complex and is obtained from what is known as a Bessel function. In practice one obtains the number of sidebands from tables or a graph. Table 1.1 gives this relationship for a few values of the modulation index.

TABLE 1.1

Number of sidebands corresponding to values of the modulation index.

MODULATION INDEX	TOTAL NUMBER OF SIDEBANDS
0·5	4
1·0	6
2	8
5	16
10	28
20	50

It will be seen from this table that for a modulation index of 5 the total number of sidebands is 16, 8 each side of the carrier. All of these sidebands are spaced 15 kHz apart (the modulating frequency) so that the total band occupied is $16 \times 15 = 240$ kHz. The sidebands are shown in figure 1.5.

If a lower modulating frequency is considered then the modulation index will be increased. Suppose that the modulating frequency is halved to 7·5 kHz, then the modulation index (assuming the same deviation) is

$$\frac{75,000}{7,500} = 10$$

From Table 1.1 it is seen that the number of significant sidebands is 28 but these will be spaced 7·5 kHz apart (the modulating frequency). Thus the total bandwidth is $28 \times 7·5 = 210$ kHz as shown in figure 1.5. This is slightly less than previously. If other modulating frequencies are taken then it can be shown that the bandwidth remains approximately the same but decreasing for low modulating frequencies. In figure 1.5 it will be seen that the amplitudes of the various sidebands vary in a random manner, and as the modulation index changes, the magnitude of any sideband varies, but not in any simple manner. It will be seen also that the magnitude of the carrier varies with the modulation index and it does become zero for certain values of the modulation index. This is quite different to the case of amplitude modulation, where the carrier remains constant. When a carrier is amplitude modulated, sidebands are produced and, since the carrier amplitude is constant, greater power must be produced by the transmitter to account for the power in the sidebands. In fact, at 100% modulation the power output is increased by 50%. In frequency modulation, since the carrier is only varied in frequency, there can be no increase in power output from the transmitter. Thus when sidebands are produced the carrier amplitude must decrease to maintain a constant power.

If the magnitude of the modulating voltage is reduced (*i.e.* the percentage modulation is reduced, the 75 kHz deviation being that at 100% modulation) then the modulation index is reduced and the number of sidebands is also reduced. Thus the bandwidth occupied by an f.m. signal depends on the percentage modulation. In amplitude modulation the bandwidth is settled by

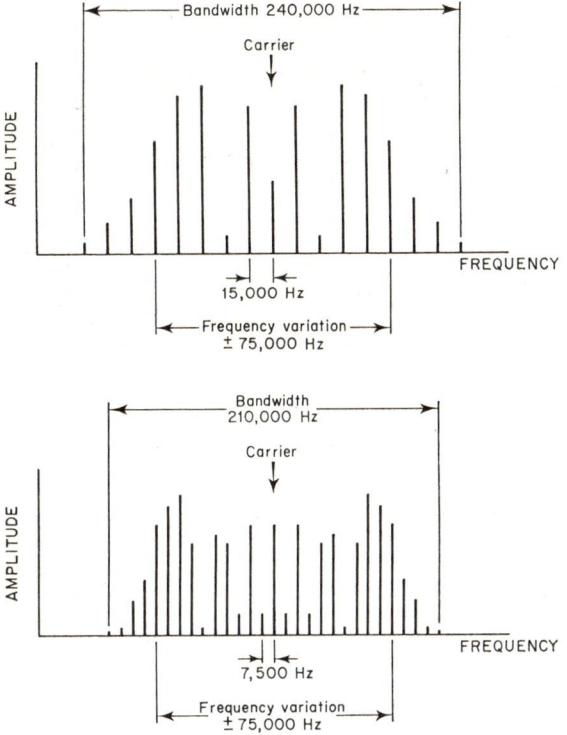

FIG. 1.5. SIDEBANDS FOR TWO DIFFERENT VALUES OF MODULATING VOLTAGE

the modulating frequency (*i.e.* the bandwidth is twice the modulating frequency). The only effect of changing the percentage modulation is to change the MAGNITUDE of the sidebands but not to change the bandwidth. This difference is rather important. It is seen that the bandwidth required for the f.m. radio transmissions is large, being about 240 kHz, compared with, say, 30 kHz for a.m. (using the same maximum modulating frequency). [The deviation does not have, of course, to be ±75 kHz and narrow band f.m. systems using reduced deviation are used for commercial purposes, but these will not be considered in this book]. The large bandwidth prevents its use on medium and long wavebands, and hence the use of v.h.f. (very high frequencies) for the f.m. transmissions.

A block diagram of a simple f.m. transmitter is given in figure 1.6. The modulating voltage obtained from the microphone is amplified by a.f. amplifier A. The carrier frequency is produced by the r.f. oscillator B. The frequency of this oscillator must be changed (*i.e.* modulated) by the audio frequency voltage from A. The output of the oscillator must now be increased in amplitude in the r.f. amplifier C and then fed to the transmitting aerial. In practice an f.m. transmitter is considerably more complex than that shown, but the principle remains the same.

It might be mentioned that instead of varying the frequency of the carrier its phase (relative to a reference) may be changed in proportion to the modulating voltage. The result is very similar to frequency modulation and, in fact, it can be shown that phase modulation gives the same result as frequency

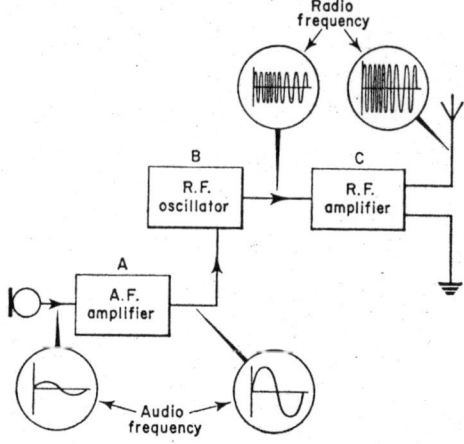

FIG. 1.6. BLOCK DIAGRAM OF A SIMPLE FREQUENCY MODULATED TRANSMITTER

modulation provided the magnitude of frequency modulation is made proportional to the modulating frequency. Phase modulation may be used in place of frequency modulation at the transmitter, the necessary compensation being made to the modulating voltage.

The basic f.m. receiver is shown in figure 1.7. In practice a superhet receiver is used, in which case the r.f. amplifier becomes the r.f. amplifier,

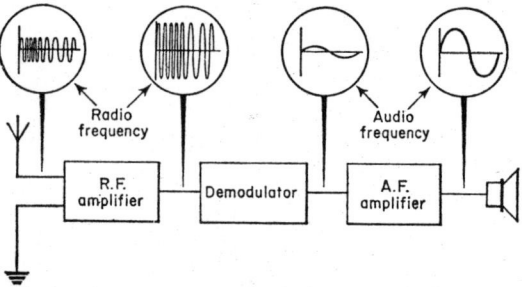

FIG. 1.7. BLOCK DIAGRAM OF A SIMPLE BASIC FREQUENCY MODULATED RECEIVER

frequency-changer and intermediate frequency amplifier. However, the basic purpose is the same—namely, increasing the amplitude of the carrier to a value large enough to operate the demodulator and also to select the required station. The demodulator serves the same purpose as in an a.m. receiver, *i.e.* to extract the modulation from the carrier. However, the circuit and method of operation of the demodulator are quite different. The demodulator is followed by an a.f. amplifier to increase the a.f. to a value that will operate a loudspeaker.

We will now consider the advantages of using f.m. in place of a.m. The main reason for using f.m. is to produce a signal with a better signal-to-noise ratio. In any system some noise is always produced, externally to the receiver or internally. This noise is produced in various ways, but consists of voltages covering the whole range of audio frequencies (and outside, but not important, in a sound receiver) and result in a background to any required signal. Ideally there should be no noise, but as there is always some, various methods are used to improve the ratio of the signal voltage to the noise voltage. This is known as the signal/noise ratio and should be as large as possible. F.M. transmissions, as used for domestic radio, produce less noise than the a.m. transmissions on medium and long waves. This is really not a fair comparison because the a.m. transmissions are narrow band (*i.e.* occupying only a small bandwidth of, say, 15 kHz), whereas the f.m. transmissions are wide band (*i.e.* they occupy a large bandwidth of, say, 240 kHz). It would be possible to use a wideband a.m. system on v.h.f., but it can be shown that the signal/ noise ratio would not be as good as with the f.m. system. Similarly, the f.m. transmission could have a smaller bandwidth by reducing the maximum deviation. However, this would reduce the signal/noise ratio compared with the wideband system. In other words the deviation and bandwidth used in f.m. is a compromise between good signal/noise ratio and excessive band-width. The figure of ± 75 kHz has been chosen as a good compromise between these two factors.

It is not easy to show that an f.m. transmission of a given bandwidth gives a better signal-to-noise ratio than an a.m. system, but a brief description follows. An interfering signal causes the carrier to be amplitude and frequency (or phase) modulated. In an a.m. receiver this amplitude modulation is demo-dulated by the demodulator (or detector) and results in noise. In an f.m. receiver the effect of the resulting amplitude modulation is removed by making the receiver or demodulator so that it only responds to frequency modulation and not to amplitude modulation. For a constant-amplitude interfering signal the amount of amplitude modulation is constant independent of frequency, and hence the noise produced is constant throughout the audio frequency band. In f.m. it can be shown that the resulting frequency modulation, produced by a constant-amplitude interfering signal, is proportional to the resulting frequency. Hence the resulting noise increases proportional to frequency. If the maximum is the same as with a.m. the mean is therefore less—ideally half. By using a relatively large deviation (*i.e.* greater than the maximum modulating frequency) a large proportion of the noise output from the demodulator is outside the audio frequency range. This improvement is proportional to the deviation since the greater the deviation the more the output will be outside the audio frequency range. Hence the use of a relatively large deviation of ± 75 kHz. Theoretically, this gives an improvement of 5 times compared with a deviation of ± 15 kHz.

To improve the signal-to-noise ratio further pre-emphasis and de-emphasis are used. These could be used on a.m. but not so effectively because the noise is produced approximately uniformly throughout the audio range. In f.m. it increases with frequency and is greatest at the extreme high frequency end of the audio range (say 15 kHz). The idea is based on the fact that, in general, the magnitude of the high frequency voltages produced by music or speech are small compared with those at low frequencies. There may be cases where this is not true, but they are rare. Thus, at the transmitter, we increase the magnitude of the high frequencies of the modulating voltage, this being called **pre-emphasis**. So that distortion is not produced, exactly the opposite is done at the receiver, called **de-emphasis**. These are done by means of a simple R-C filter, that at the receiver having the opposite characteristic to that at the transmitter. Since the high frequencies are reduced at the receiver, so are the noise voltages and, since the noise voltages are greater at high frequencies, a considerable improvement in signal/noise ratio is obtained. In practice the improvements obtained are somewhat reduced owing to the effect of multiple frequency noise, but there is no doubt that f.m. produces a considerably better signal/noise ratio than the corresponding a.m.

The signal/noise ratio is important if one considers the dynamic range of a signal. The dynamic range is the range between the maximum power of the audio frequency signal and the smallest power. In an orchestra this ratio may be as high as 70 dB (10 million to 1). In an a.m. signal the noise voltage may be, say, 5% of the signal with maximum (100%) modulation. Thus, when the signal produces only 5% modulation it becomes lost in the noise. Therefore, the voltage ratio of the signal is 100/5, or, in terms of power ratio (power being proportional to voltage squared), it is

$$\left(\frac{100}{5}\right)^2 = 20^2 = 400 \text{ or approximately 27 dB.}$$

Thus on an a.m. transmission the power variations transmitted must be reduced to approximately this ratio by what is known as **volume compression**. Some of the qualities of the music are thereby lost. By using f.m. the compression need not be so high, and a ratio of 50 dB or so can be transmitted.

V.H.F. TRANSMISSIONS

The BBC v.h.f. transmissions in this country cover a range of frequencies from about 88 MHz to 98 MHz. Practically all receivers, however, tune from about 88 MHz to 106 or 108 MHz. The deviation corresponding to 100% modulation is ±75 kHz and the maximum audio frequency transmitted is 15 kHz. This means that the bandwidth used by each transmitter is 200-240 kHz at 100% modulation. For most of the time the percentage modulation will be much less, and so the bandwidth used will also be much less. Stations have frequency separations of 100 kHz, but, of course, two stations so close as regards frequency are a long distance apart geographically. The three BBC transmissions in any area are about 2 MHz apart.

CHAPTER 2

THE F.M. RECEIVER

I N this chapter we will deal with the operation and circuits of the various parts of the receiver. Only a superhet will be considered as all commercial receivers are of this type.

Before dealing with the f.m. receiver we will consider the performance criteria as these are more involved than in an a.m. receiver. In order that the performance of one receiver may be compared with another it is important to have standard tests and standard ways of expressing the performance. If different tests are done by various manufacturers it becomes almost impossible to compare the performance of one receiver with that made by another manufacturer. There are a number of standards throughout the world, but the two most commonly used and commonly quoted are the German D.I.N. standard (Deutsche Industrie Norman) and the American I.H.F. standard (Institute of High Fidelity Manufacturers of America). The standards are quite different: the D.I.N. specification lays down MINIMUM standards that equipment must reach; of course, many receivers exceed the D.I.N. specification. The I.H.F. specification only describes how the tests shall be performed and how the results shall be presented, and does not lay down any minimum standards. The specifications are quite complex and only simple explanations will be given in this book: those wishing to do the tests should consult the appropriate literature. We will now consider some of the important properties of f.m. receivers.

SENSITIVITY

The sensitivity of an f.m. receiver has little meaning and must be related to the signal/noise ratio. As the signal input is reduced to a low value the signal output will reduce until at some point the noise is such that the result is not acceptable. Hence the term "usable sensitivity" is often quoted, which is the magnitude of the input signal to give a certain signal/noise ratio, usually 30 dB. The carrier must be frequency modulated with an a.f. signal, commonly 400 or 1,000 Hz, and the percentage modulation or deviation must be quoted. This may be 100% ($\pm$75 kHz) or 30%.

If the input to an f.m. receiver is varied and the output measured a result similar to figure 2.1 is obtained. Starting at point A, as the input increases the output increases, initially the rate of change of output will be approximately equal to the rate of change of input, but limiting will occur (f.m. receivers have limiters to reduce the effects of a.m.), and beyond point B the output is almost constant and independent of the input. The signal/noise ratio is also shown, which increases as the input is increased from A to B, but after B the ratio becomes almost constant.

Thus the usable sensitivity may be quoted as the input corresponding to C, where the signal/noise ratio is 30 dB. High-class tuners may go down to a usable sensitivity of 1 to 2 μV. What may be more important is the ultimate signal/noise ratio for larger signals, particularly if the receiver is to be used on stereo. For a good receiver this should be, say, 70 dB on a mono signal.

CAPTURE RATIO

F.M. receivers have the ability to reject a weaker station when a more powerful station is received at approximately the same frequency. This is known as **capture effect**. I.H.F. quote this measurement in terms of the ratio of the two signals when the interference produced by the unwanted station is -30 dB of the signal from the wanted station. The ratio may be 6 dB for poor receivers and as low as 1 dB for high-class tuners.

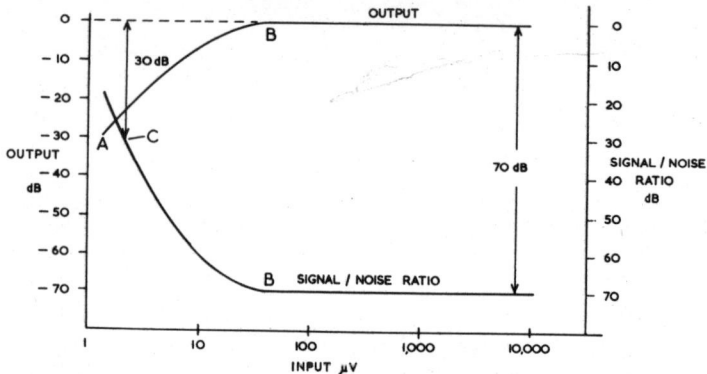

FIG. 2.1. OUTPUT AND SIGNAL/NOISE RATIO AGAINST INPUT VOLTAGE FOR A TYPICAL
F.M. RECEIVER

SELECTIVITY

This is given by I.H.F. as the ratio of wanted to unwanted signal inputs to produce a disturbance of −30 dB due to the unwanted signal on the wanted signal. This may be quoted for adjacent channels (200 kHz) or alternate channels (400 kHz). The ratio is expressed in dB and may be 40 dB for a cheap receiver to 75-80 dB for a good tuner, both figures being for alternate channels.

A.M. REJECTION

It has already been mentioned that an f.m. receiver should reject an a.m. signal. This is quoted as a ratio in dB of the outputs due to equal f.m. and a.m. signals. It should be 50-65 dB.

IMAGE REJECTION

This is the rejection to an input at the image frequency (*i.e.* one at twice the i.f. from the wanted signal) and expressed in dB. Should be 45 to 90 dB, the higher figure being for good tuner units.

I.F. REJECTION

This should be 90-100 dB.

F.M. RECEIVER

A block diagram of a superhet f.m. receiver is shown in figure 2.2.

The signal from the aerial is amplified by the r.f. amplifier and then fed to the mixer, which is fed from the oscillator. If the incoming frequency is f_i and the oscillator frequency is f_o then, from the mixer, we get (together with other frequencies) the intermediate frequency $f_o - f_i$. This is amplified by the i.f. amplifier and then fed to the f.m. demodulator. The a.f. output of the demodulator is then amplified in the audio frequency amplifier and fed to the loudspeaker.

The various blocks will now be considered. The circuits used depend on the results to be obtained. F.M. equipment ranges from portable receivers at a low price to what are commonly termed "tuner units" at high prices. The tuner unit does not include the a.f. amplifier as it is intended to feed a hi-fi amplifier. In both extreme cases the principles are the same, but different circuits are used in tuner units to obtain better performance. Generally, the greater the cost of the equipment, the more sophisticated the circuits and the

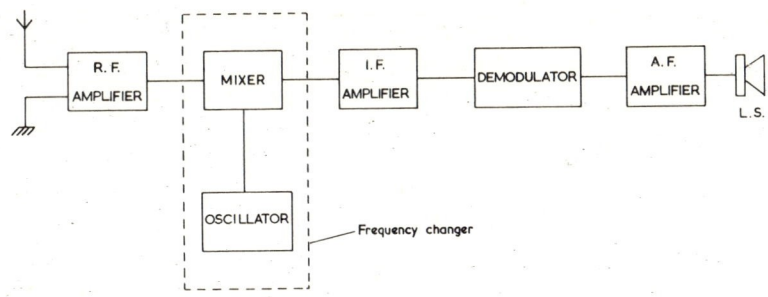

FIG. 2.2. BLOCK DIAGRAM OF F.M. RECEIVER

better the performance. Both extremes will be considered but, of course, there are many sets falling in between them.

R.F. AMPLIFIERS

The r.f. amplifier serves three purposes:

(a) It acts as an amplifier and increases the magnitude of the signal applied to the mixer of the frequency changer stage. All stages produce some noise, and a frequency-changer produces more noise than an amplifier. It is therefore desirable to increase the magnitude of the small signal on the aerial before it is applied to the frequency-changer, so that it is large relative to the noise produced by this stage. This results in a receiver having an improved signal/noise ratio, particularly on weak signals.

(b) It acts as a preselector stage, in particular, to remove the second channel interfering signal. Suppose that the input frequency is f_i; the oscillator frequency is f_o; and the intermediate frequency is f_{if}. Further, we will assume that the oscillator frequency is higher than the input frequency, as this is almost always the case. For the wanted station then the frequencies are such that $f_{if} = f_o - f_i$. However, there is another frequency which is the i.f. above f_o, which will also produce the same intermediate frequency. For example, suppose $f_{if} = 10$ MHz and $f_i = 100$ MHz then the oscillator frequency is 110 MHz, so that the i.f. is $110 - 100 = 10$ MHz. Now a second station of frequency 120 MHz also gives a difference-frequency of 10 MHz (*i.e.* $120 - 110 = 10$ MHz). This is known as the second channel station or second channel interference. Once these two incoming frequencies have been changed to the i.f. there is, of course, no way in which they can be separated. Hence, it is essential to remove the second channel station before it reaches the frequency-changer or mixer. Thus, the r.f. amplifier must have sufficient selectivity to do this. The second channel station is, as seen, twice the intermediate frequency from the wanted station. If a signal of the intermediate frequency reaches the frequency-changer it will pass straight through it and into the i.f. amplifier. Therefore it is essential for the r.f. amplifier to remove any input signal at the intermediate frequency.

(c) If the aerial is connected directly to the frequency-changer, it is possible for the oscillator to feed into the aerial and cause radiation at the oscillator frequency. This is likely to cause interference with other receivers. By using an r.f. amplifier stage between the frequency-changer and aerial most of this possible radiation will be removed, because an amplifier will only operate in one direction, *i.e.* from input to output. In practice, due to capacitances, there may be a slight feed in the other direction, but this can be made very small. Thus the r.f. amplifier acts as a buffer stage between oscillator and aerial.

We will first consider the normal small portable f.m. receiver. In these receivers the r.f. stage is normally a single stage using a bipolar transistor. This is often operated as a common base circuit since the frequency response for a given transistor is better using this connection. A common emitter circuit may be used, but to get amplification at the high frequency of approximately 100 MHz a "better" transistor is required, *i.e.* a transistor with a higher cut-off frequency, f_T, must be used. When a common base circuit is used, since the input impedance is low (approximately 50 Ω), the input tuned circuit is heavily damped and has limited selectivity. As the range of frequencies over which the receiver must tune is small there is no point in making this tuned circuit variable. A typical circuit is given in figure 2.3. The input tuned circuit is L_2 C_1, the circuit being completed through C_2 which acts as a d.c.

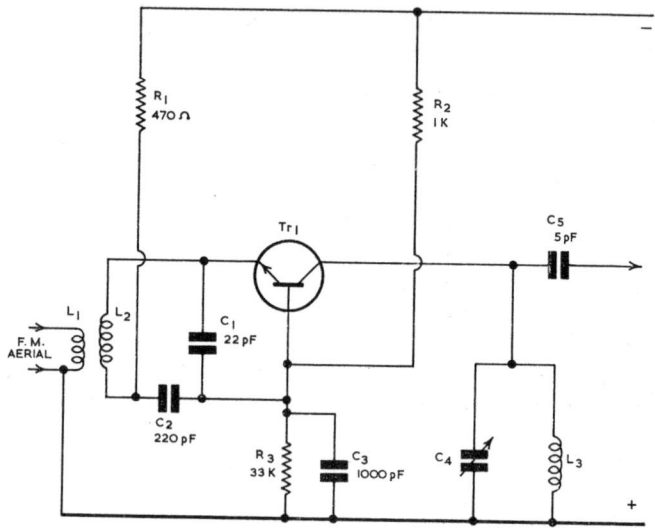

FIG. 2.3. R.F. AMPLIFIER OF SMALL PORTABLE F.M. RECEIVER USING A COMMON BASE CIRCUIT

blocking capacitor. The aerial is coupled to this tuned circuit by the coupling winding L_1. The output tuned circuit is L_3 C_4 and is tuned by C_4, which is ganged to the oscillator tuning capacitor. It will be seen that the collector is connected to the earthy line through the tuned circuit. In order to explain the d.c. operating conditions of the circuit it has been redrawn in figure 2.4. The potential divider R_2 R_3 provides a suitable base voltage and the base voltage will be almost equal to the supply voltage (say 1 volt less) by, as in this case, making R_3 much greater than R_2. The emitter is returned to negative line through L_2 and R_1, thus the emitter current flows in R_1 and forms the normal emitter stabilizing resistor. A d.c. negative feedback voltage is produced across the resistor which reduces variations in emitter current due to temperature, etc. The base is connected to the earthy line (+ve) as far as a.c. is concerned by C_3, and the lower end of L_2 and C_1 are also connected to the positive line through C_3 and C_2. Thus R_1 is by-passed by C_3 (and C_2) so that the gain is not reduced due to a.c. negative feedback. R_1 is, of course, of low value, the voltage across it normally being about 1 volt. This type of connection is commonly used and is perhaps more difficult to understand than

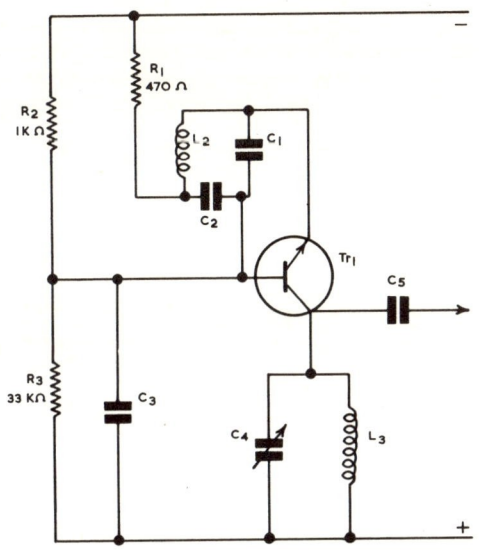

FIG. 2.4. CIRCUIT OF FIGURE 2.3. REDRAWN

the circuit where the collector is returned to the upper supply line. It has the advantage that one end of C_4 is at earthy potential. The voltage across the output tuned circuit L_3 C_4 is fed to the mixer through the small capacitor C_5.

Another circuit is given in figure 2.5, but in this case a common emitter stage has been used. The input impedance is now much higher, but the input tuned circuit is still not made variable, although a tap is used on the tuned circuit to reduce the loading (this could have been used in figure 2.3). The

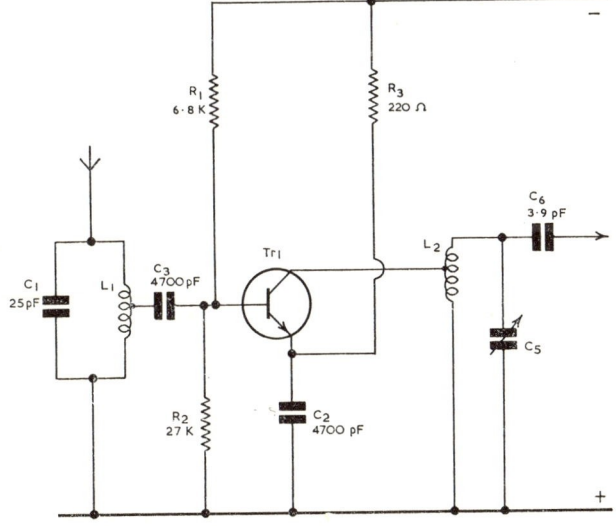

FIG. 2.5. R.F. AMPLIFIER OF SMALL PORTABLE RECEIVER USING COMMON-EMITTER CIRCUIT

input tuned circuit is L_1 C_1 and the aerial (commonly the rod type of aerial) is connected directly to this tuned circuit. This circuit is again the upside-down type of circuit with the collector taken to the earthy line (positive), through the tuned circuit L_2 C_5, while the emitter is taken through the stabilizing resistor R_3 to the negative line. The output tuned circuit is tuned by C_5, ganged to the oscillator tuning capacitor. Bias is provided by R_1 and R_2; C_3 is a d.c. blocking capacitor to prevent the bias being shorted out by the low resistance d.c. circuit through L_1. R_3 is effectively by-passed by C_2 (or one may consider that the emitter is connected to the earthy line as regards a.c.). A tap is used on the output tuned circuit L_2 C_5 to reduce the damping produced by the transistor, and the output is taken to the mixer stage through C_6. The performance of tuner units is generally superior, since they are usually more expensive and better arrangements can be used. One problem that may arise in an r.f. stage is what is called **cross-modulation.** If two modulated signals are fed into a non-linear device one signal may become modulated by the other. This is most likely to occur if one signal is large, in which case the modulation of the larger is likely to become impressed or modulated on to the weaker. Once this has occurred no amount of selectivity will remove the effect of the larger signal on the smaller. A way to reduce this effect is to avoid a large unwanted signal being fed to the r.f. amplifier and to make the r.f. amplifier as linear as possible. The use of a field effect transistor (F.E.T.) has advantages compared with the use of a bipolar transistor. Firstly, the input impedance of an F.E.T. is very high compared with that of a bipolar transistor. This means that the damping of the first tuned circuit is greatly reduced; it therefore has a greater selectivity and will probably be tuned, so largely rejecting any unwanted station. Secondly, the characteristics of an F.E.T. are more linear. This perhaps needs some explanation. In figure 2.6(a) is shown the $I_c - V_{be}$ characteristic of a typical bipolar transistor which has a reasonably linear characteristic from, say, C to D. However, around part B the characteristic is very non-linear, and little current flows over the portion A-B. Thus, if a bipolar transistor is overloaded (by a powerful signal) it operates over this very non-linear portion and cross-modulation is likely to

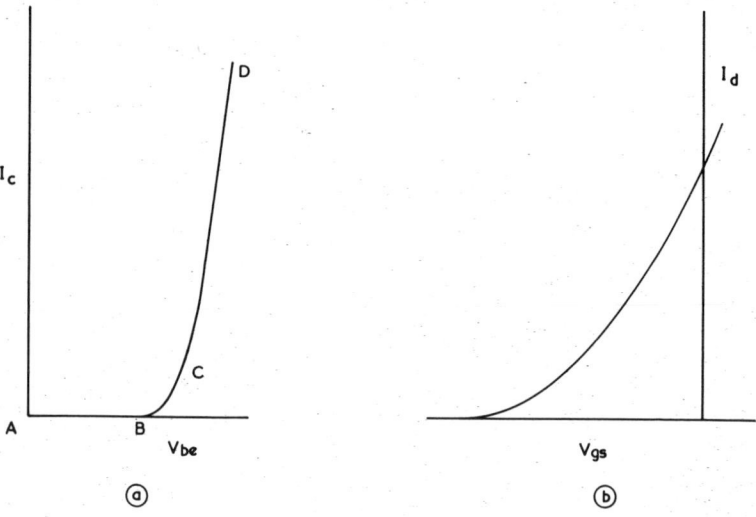

FIG. 2.6. CHARACTERISTICS OF (a) BIPOLAR TRANSISTOR: AND (b) FIELD EFFECT TRANSISTOR

occur. At (b) is shown the I_d—V_{gs} (drain current-gate-source voltage) characteristic of an F.E.T. The characteristic is curved throughout and, in fact,
approximates to a square law characteristic, but at no point do we get the
severe non-linearity of the bipolar transistor. Hence cross-modulation is less
likely with an F.E.T. The F.E.T. may be connected in a common source or
common gate circuit; an example of a common source circuit is given in
figure 2.7. The input impedance of such a circuit is very high and hence there
is little damping of the input tuned circuit L_2 C_1. L_1 is the coupling winding

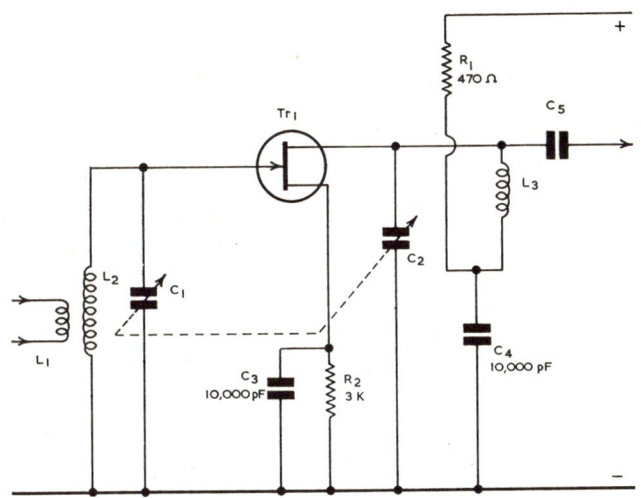

FIG. 2.7. R.F. AMPLIFIER USING F.E.T. IN COMMON-SOURCE CIRCUIT

from the aerial. Owing to low damping and high selectivity the circuit is tuned
to the incoming frequency by making C_1 variable, this capacitor being ganged
to C_2 and oscillator tuning capacitor. If there is now a large unwanted signal
on the aerial it will greatly be reduced relative to the wanted station by the
high selectivity of this tuned circuit before it reaches the transistor. An F.E.T.
is normally conducting, and requires a negative bias (for an n-channel F.E.T.
as shown), which is obtained by the resistor R_2 placed in the source lead. To
prevent a.c. negative feedback and reduction of gain this is by-passed by C_3.
The output tuned circuit is L_3 C_2, L_3 being by-passed to the earthy supply
lead by C_4, which forms a decoupling circuit with R_1. With this circuit there
is some capacitance between drain (the output connection) and the gate (the
input connection). Although it may only be a few picofarads the circuit may
oscillate due to the small damping on both input and output circuits.

This may be overcome by the use of a common gate circuit, an example
of which is given in figure 2.8. In this case the vital feedback capacitance is
that between drain and source, and is extremely small. The input impedance
of this circuit is, however, much lower than the common source circuit, but
still much higher than that of a bipolar circuit. Thus the input tuned circuit
L_2 C_1 still gives good selectivity and is tuned by making C_1 variable. The
output tuned circuit is L_3 C_2, L_3 being connected to the earthy lead by C_3,
which forms part of the decoupling circuit R_1 C_3. As this circuit is drawn there
is no bias on the F.E.T. However, no appreciable gate current flows in an
F.E.T. until the gate is about 0·5 volt positive relative to the source, and hence
the circuit can be satisfactory without bias. Self-bias may be added to deal
with larger signals. This is obtained by connecting the parallel combination

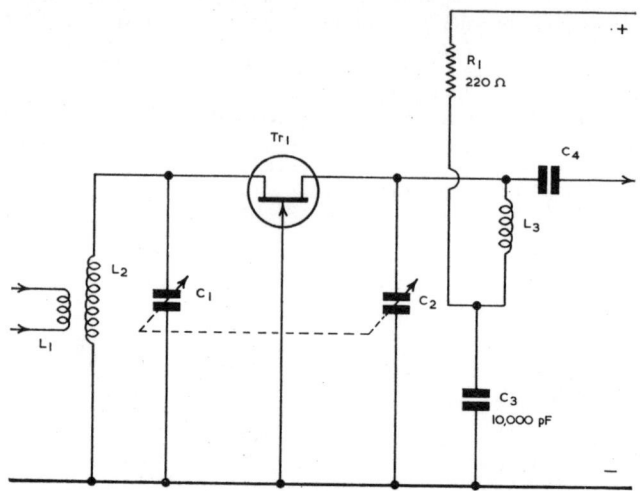

FIG. 2.8. R.F. AMPLIFIER USING F.E.T. IN COMMON-GATE CIRCUIT

of a resistor of, say, 2 MΩ and a capacitor of, say, 10,000 pF in series with the gate. If gate current does flow then the capacitor becomes charged and provides some negative bias on the gate.

The best of both circuits (*i.e.* high input impedance and good stability) can be obtained by a combination of the two circuits, called a **cascode circuit.** A typical circuit is given in figure 2.9. In this case two F.E.T.s are used, connected in series as regards the d.c. supply, the source current of Tr_2 being the drain current of Tr_1. Tr_1 is connected as a common-source amplifier, since

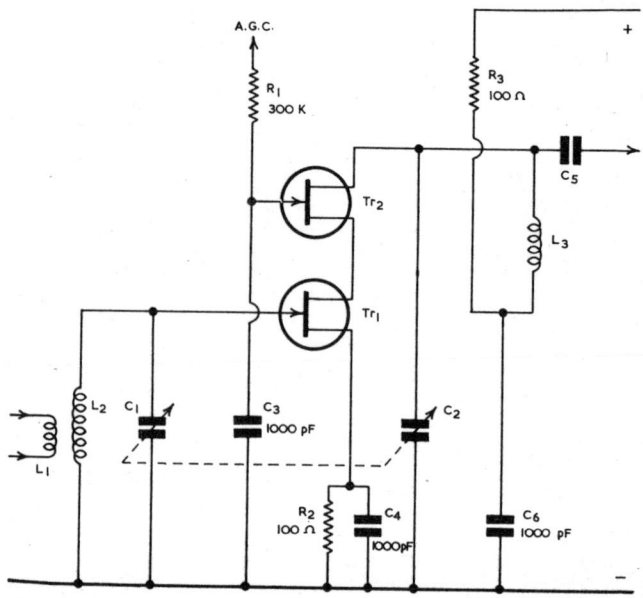

FIG. 2.9. CASCODE R.F. AMPLIFIER USING FIELD EFFECT TRANSISTORS

the source is by-passed to the earthy line by C_4, while Tr_2 is connected as a common-gate circuit since the gate is by-passed to the earthy line by C_3. Bias for Tr_1 is provided by R_2. $L_2 C_1$ form the input tuned circuit and since the damping produced by Tr_1 will be very low this circuit, tuned by C_1, will have good selectivity. A suitable positive voltage must be fed to the gate of Tr_2; this could be obtained by a suitable potential divider circuit across the supply. In this particular circuit an a.g.c. voltage is fed to the gate through R_1, so the gain can be varied. The use of a.g.c. will be considered later. The voltage fed to the gate is, in fact, a positive voltage from a potential divider, but the voltage is reduced on a large signal by a negative a.g.c. voltage opposing the positive voltage. The output tuned circuit is $L_3 C_2$ and $R_3 C_6$ is decoupling.

A dual-gate n-channel, depletion type, MOSFET (metal oxide semiconductor field effect transistor) may be used as an r.f. amplifier, and a basic circuit is given in figure 2.10. The input impedance of a MOSFET is very high (the gate current is about 50 nA or 50×10^{-9} A, max.), and hence a variable

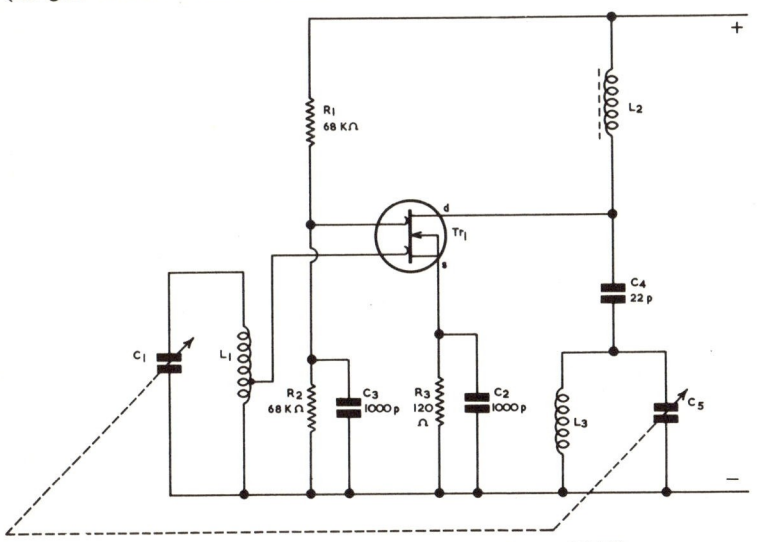

FIG. 2.10. R.F. AMPLIFIER USING DUAL GATE MOSFET

input tuned circuit $C_1 L_1$ is used. R_3 in the source circuit provides bias and is by-passed by C_2 to prevent negative feedback. The second gate is given a positive voltage from the potential divider $R_1 R_2$ and is by-passed to the earthy line by C_3. The drain is fed with a positive voltage through the inductor L_2, and the output tuned circuit $L_3 C_5$ is fed through d.c. blocking capacitor C_4. The output may be taken directly to another MOSFET operated as a mixer (see figure 2.18).

FREQUENCY-CHANGERS

The purpose of the frequency-changer is to change the variable incoming frequency from the r.f. amplifier to a fixed intermediate frequency, which is fed to the i.f. amplifier. The incoming frequency range is 88-98 MHz, although most receivers tune from 88 to 108 MHz. The i.f. used almost internationally is 10·7 MHz. The frequency-changer consists of two sections: an oscillator and a mixer. The oscillator produces the required local frequency and this is "mixed" in the mixer with the incoming frequency to produce a difference-frequency equal to the i.f. The oscillator may be the intermediate frequency below or above the incoming frequency, but, in most f.m. receivers, the oscillator frequency used is that above the incoming frequency.

The mixing process may be additive or multiplicative. In the additive type the two signals are added together and then fed to some non-linear device which then generates the difference-frequency (it also generates the sum frequency, which is not used). It is important to note that just adding the two frequencies together does NOT produce a difference-frequency—a non-linear device is essential. In multiplicative mixing the two signals are multiplied together, so producing the difference-frequency. Multiplicative mixing was used with valves, but transistors normally use the additive process.

First we will consider the type of circuit used in small portable receivers, where cost and simplicity form probably the most important factor. A bipolar transistor is normally used, often operating as an oscillator and mixer. A typical circuit is given in figure 2.11 where the transistor Tr_1 is connected in a common-base circuit since the base is by-passed to the earthy supply lead

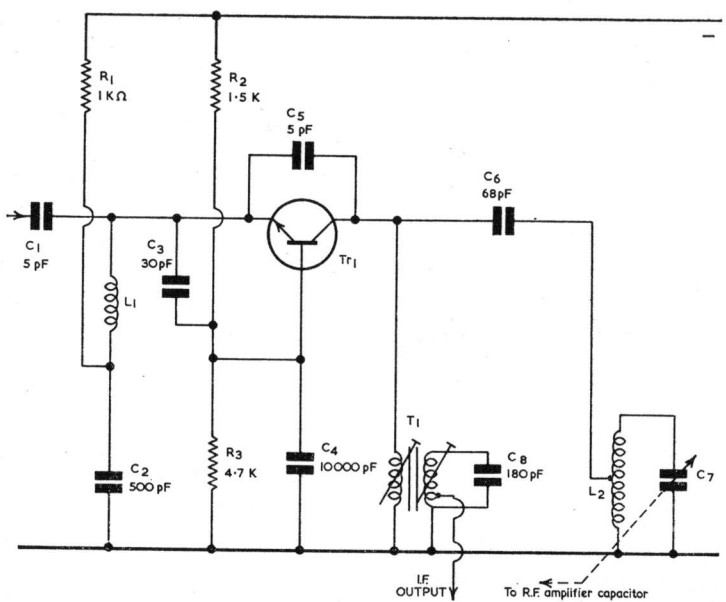

FIG. 2.11 MIXER-OSCILLATOR STAGE

by C_4. The incoming signal is fed through the small capacitance C_1. The reactance of this capacitor reduces any loading, by the low input impedance of the transistor, on the tuned output circuit of the r.f. stage. It also reduces the possibility of oscillator radiation from the aerial and pulling of the oscillator frequency by the r.f. stage. The emitter is returned to the negative line through L_1 and R_1 (the emitter stabilizing resistor). As regards a.c. L_1 is returned to the earthy line through the low reactance of C_2. L_1 will have a high impedance at the incoming and oscillator frequencies, and so have little effect. However, due to the non-linearity of the transistor, mixing of the incoming and oscillator frequencies occurs in the emitter circuit with the production of the intermediate frequency (having a value of about one-tenth of the input frequency). Hence, L_1 provides a relatively low impedance path (L_1 may resonate with C_2 at the intermediate frequency) for the flow of the i.f. current, which in turn, causing a corresponding collector current to flow.

This circuit also acts as an i.f. trap rejecting any i.f. picked up by the aerial. The tuned circuit, L_2 C_7, which determines the frequency of oscillation, is in the collector circuit. A tap is used on L_2 to reduce any damping on the tuned circuit so that its Q is high, resulting in good frequency stability. Since the transistor is connected as a common-base circuit there is no phase reversal between input (emitter) and output (collector) circuits. The circuit can therefore be made to oscillate by the use of the small capacitor C_5 connected between collector and emitter. C_5 and C_3 form a potential divider so that the emitter is fed with a lower voltage than the collector. One can re-draw this circuit, as in figure 2.12, when it appears as a Colpitt's circuit, C_3 and C_5 forming the two capacitors of the tuned circuit, the inductance of which is

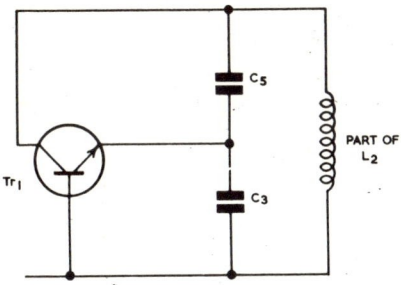

FIG. 2.12. OSCILLATOR PART OF FIGURE 2.11 REDRAWN

part of L_2. As a result of Tr_1 operating as an oscillator and the input signal being fed through C_1 there are two voltages of different frequencies on the emitter, one of oscillator frequency and one of the incoming frequency. Due to the non-linearity of the transistor the difference-frequency is produced in the collector, this being the i.f. output. Thus the mixing process is one of additive mixing, *i.e.* adding the two signals and passing them through a non-linear device. There are many frequencies in the collector circuit but the two most important are the oscillator frequency and the i.f. The i.f. frequency is picked out by the i.f. tuned transformer T_1, which is tuned to 10·7 MHz. The output is taken from a tap on the secondary, so that the Q of the secondary circuit is maintained at a high value. The value of C_6 is of some importance. If it were not present then the i.f. would tend to be shorted out by L_2, since its reactance would be low at this relatively low frequency. C_6 is therefore chosen to have a relatively high reactance at the i.f. so that the shorting effect of L_2 is reduced. It must, however, have a low reactance at the oscillator frequency so as to couple the transistor to the tuned circuit L_2 C_7. The tuning capacitor C_7 is ganged with those in the r.f. amplifier. The collector is connected to the earthy supply as regards d.c. through T_1. Bias for the base is provided by R_2 R_3.

Another circuit is given in figure 2.13, where a different arrangement is used to separate the oscillator and intermediate frequency voltages. The transistor is connected common base, with emitter stabilizing resistor R_1 and bias potential divider R_2 R_3. The input frequency is fed through the small capacitor C_1 from the r.f. stage, and L_1 and C_2 form the i.f. trap as in the last circuit. Instead of placing the oscillator-tuned circuit and i.f. transformer in parallel they are now placed in series, T_1 being the i.f. transformer and L_2 C_7 the oscillator tuned circuit. Feedback to make the circuit oscillate is now by C_5 connected from the oscillator-tuned circuit to the emitter, and again C_3 C_5 form a potential divider. L_1 acts in the same way as L_1 of figure 2.11. At the oscillator frequency the reactance of C_8 will be low, hence the circuit L_2 C_7 is virtually connected to the collector of Tr_1. At the relatively low i.f. the

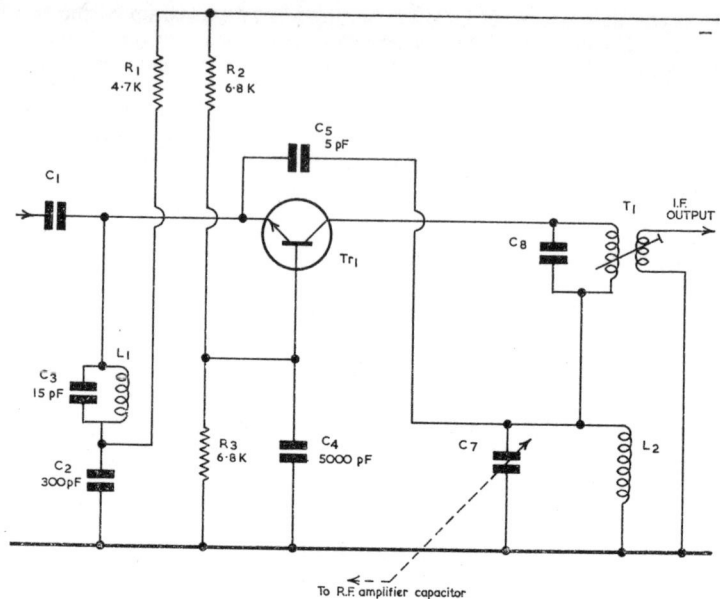

FIG. 2.13. ANOTHER MIXER-OSCILLATOR STAGE

reactance of L_2 will be low, and so the primary of T_1 is the effective load on the transistor.

The two foregoing circuits have the advantage of simplicity and of only using one transistor. However, by using a transistor for two operations at the same time makes it difficult to get ultimate performance. A circuit, still using bipolar transistors but using a separate oscillator, is shown in figure 2.14. This arrangement tends to be used in the more expensive receivers. Tr_1 forms the mixer and is connected as a common-emitter circuit with the emitter by-passed to the earthy line by C_3. R_3 is the emitter stabilizing resistor and bias is provided by R_1 R_2. As before, the input is fed through the small capacitor C_1. C_2 is simply a d.c. blocking capacitor to prevent L_1 shorting out the bias voltage across R_2. L_1, as in the last two cases, produces a low impedance circuit at i.f. and may resonate with C_2. The base is also fed with a voltage from the oscillator through C_8. The resultant i.f. is picked out by the transformer T_1. The collector circuit is now simpler because there is no oscillator-tuned circuit to accommodate. Tr_2 forms the oscillator circuit, which is now a common-base circuit, the base being connected to the earthy line through C_7. C_6 feeds voltage from collector to emitter to cause oscillation, and C_6, C_{10} make the circuit into a Colpitt's oscillator circuit. Bias is provided by R_4 R_6, and R_5 acts as the emitter stabilizing resistor. Flow of base current will probably produce some self-bias across C_7. The oscillator tuned circuit is L_2 C_9, C_9 being the tuning capacitor.

When we come to tuner units bipolar transistors may still be used in circuits similar to that of figure 2.14, or an FET may be used for the mixer. As explained in connection with the r.f. amplifier, an FET has a smoother characteristic than that of a bipolar transistor, and for this reason makes a better mixer. The fact that an FET follows an almost square law characteristic makes it ideal for a mixer. Obviously, the mixer must be non-linear, but if it follows a square law characteristic (rather than a law involving higher powers)

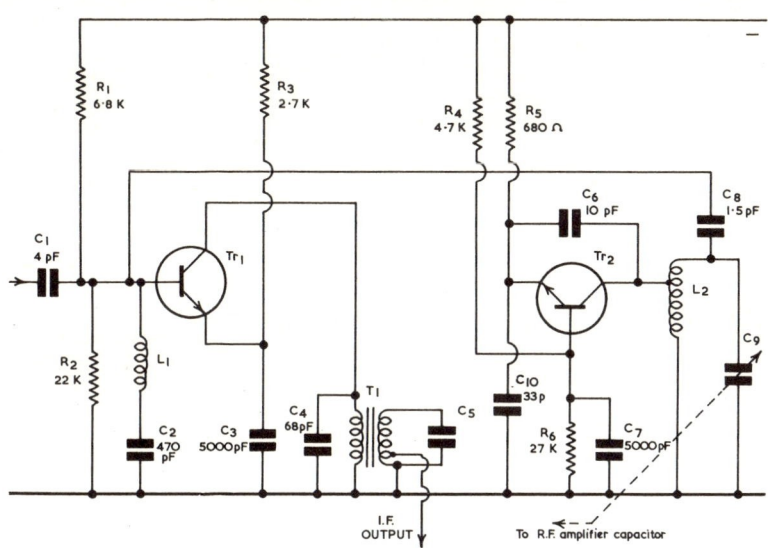

FIG. 2.14. FREQUENCY-CHANGER WITH SEPARATE OSCILLATOR

then fewer unwanted frequencies are produced in the output as a result of mixing. A circuit using an FET mixer but bipolar transistor as an oscillator is given in figure 2.15. To improve the selectivity, and prevent overloading the mixer (with possible cross-modulation) by a powerful unwanted station, three tuned circuits are used between the r.f. amplifier and mixer Tr_1. Circuits L_1C_1 and L_2C_2 are coupled by the small top capacitance C_4. Circuits L_2C_2 and L_3C_3 are coupled by mutual inductance between L_2 and L_3. These three circuits form a band-pass circuit and, if correctly designed, result in a flat response over a small band of frequencies, the response curve having steep sides as in figure 2.16. This is, of course, the ideal shape to receive the carrier and its sidebands but reject other stations outside this range of frequencies. The mixer is fed from coupling winding L_4 and will cause little damping because it is connected as an FET common-source circuit. The gate is therefore fed with the incoming signal, but the source is fed with the oscillator frequency from winding L_7. Thus, between gate and source both voltages occur and mixing results in the intermediate frequency being produced in the drain circuit, the frequency being picked out by the i.f. transformer L_5C_5 and L_6. R_1C_6 forms a decoupling circuit. Bias for Tr_1 is provided by the components R_2C_7 in the source lead.

Tr_2 forms the oscillator and is connected as a Hartley oscillator circuit using a tapped inductor L_8. The oscillator is shown in a simplified form in figure 2.17. One may consider the circuit in a number of ways. As the circuit is drawn one may consider that Tr_2 is connected as a common-collector circuit or emitter-follower. The input to the base is from the top end of L_8 and the output is taken from the emitter to the tap. There is no voltage gain in an emitter-follower, but the loop gain is greater than unity due to the voltage step-up (from emitter to base) by L_8. Thus the circuit will oscillate. Since the emitter current is much greater than the base current there is no difficulty in providing the base current, although L_8 acts as a step-up voltage transformer (and therefore step-down in current). Turning to figure 2.15, the tuned circuit is L_8C_9. the a.c. circuit of L_8 being completed through C_8. Starting bias is provided by R_4R_5. This is generally necessary with transistor oscillators due

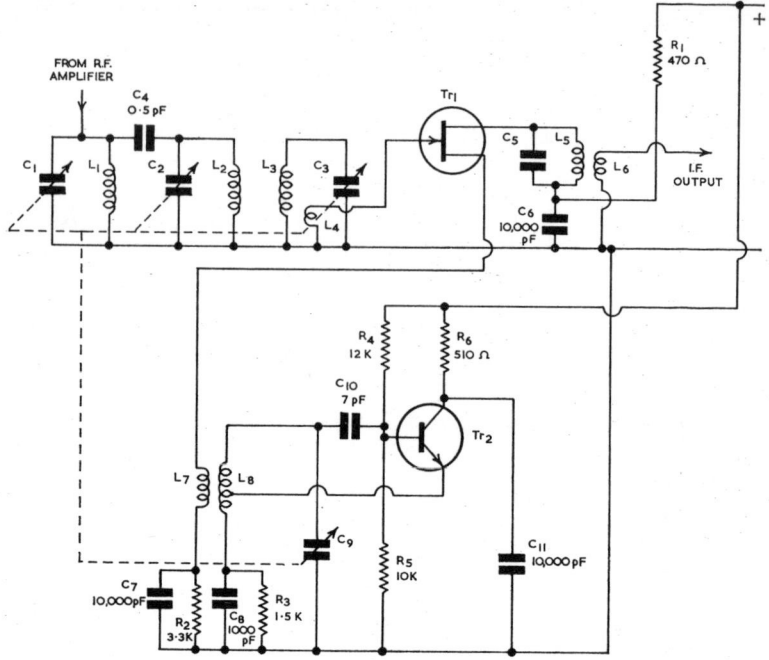

FIG. 2.15. FREQUENCY-CHANGER USING F.E.T. MIXER

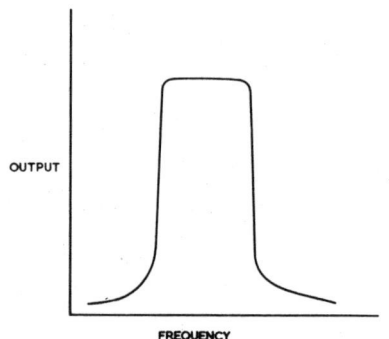

FIG. 2.16. RESPONSE CHARACTERISTIC OF BAND-PASS CIRCUIT

to the fact that the collector current of a transistor is almost zero with no base bias. Thus there is no collector current to start the oscillation process. The emitter current will flow through R_3 and will produce some self-bias due to the capacitor C_8 i.e. the pulses of emitter current (when oscillating class-C) will charge up C_8 so as to tend to reverse bias the emitter as is required for class-C operation. C_{10} prevents the bias being upset due to the circuit through L_8 to the emitter. C_{11} connects the collector to the negative earthy line as regards a.c., and completes the circuit to the lower end of L_8 (this is seen more easily in figure 2.17). Positive voltage is fed to the collector through R_6. This circuit has the advantage that the gate input impedance is high and therefore there is little damping of the input tuned circuits. However, the impedance

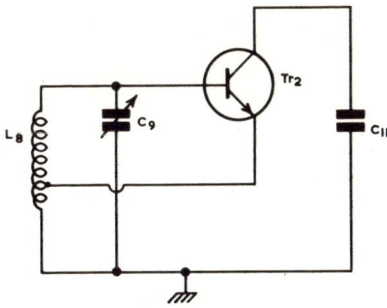

FIG. 2.17. OSCILLATOR SECTION OF FIGURE 2.15

into the source terminal is low (say 150 Ω) and considerable power is required from the oscillator. This requires tight coupling to the oscillator, which may cause oscillator-pulling by the input circuits. One alternative is to feed the input to the source but this causes serious damping of the input circuit. However, the power from the oscillator is now low and the coupling can be loose. Another way is to feed both inputs to the gate in series. This is claimed to give a higher gain and low oscillator power and loose coupling, so reducing oscillator pulling.

Another method is to use an n-channel, depletion type dual gate MOSFET and a circuit is shown in figure 2.18. The input tuned circuit $L_1 C_2$ is fed through the capacitor C_1 from the drain of the MOSFET used as the r.f. amplifier. $L_1 C_2$ corresponds to $L_3 C_5$ of figure 2.10. The first gate is fed from the tap on L_1 with the incoming frequency, and bias is provided by the drop across R_2, which is by-passed by C_4. The second gate is fed with a positive voltage from the potential divider $R_1 R_3$ and is also fed with the oscillator voltage through C_3. Mixing takes place (as a multiplication process) in the MOSFET, and the intermediate frequency is picked out by the i.f. transformer T_1, the primary

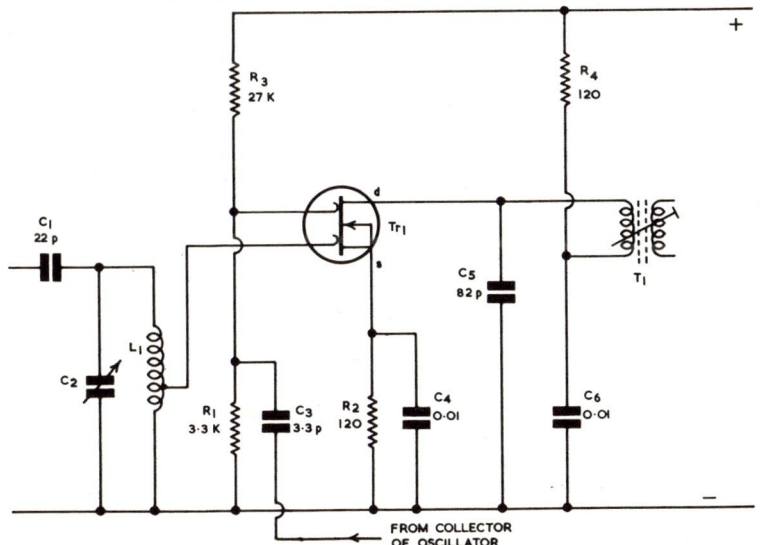

FIG. 2.18. MIXER USING DUAL GATE MOSFET

of which is tuned by C_5. R_4 and C_6 are decoupling. One advantage in this device is that there is negligible coupling between oscillator and the input circuit.

There are, of course, a large number of possible variations of the mixer, oscillator and r.f. amplifier sections. Instead of using a small capacitance to couple the r.f. amplifier to the mixer, a transformer may be used. Instead of using variable capacitances for tuning the various tuned circuits variable inductors may be used, called **permeability tuning** since the inductances are varied by varying the positions of the ferrite cores. The tuning may also be achieved by the use of variable capacitance diodes, but this is described later.

Another method of feeding the bias is shown in the basic circuit of figure 2.19. This may be an r.f. amplifier or mixer stage. $L_1\ C_1$ is the input tuned circuit and is returned to the earthy line through the emitter stabilizing resistor

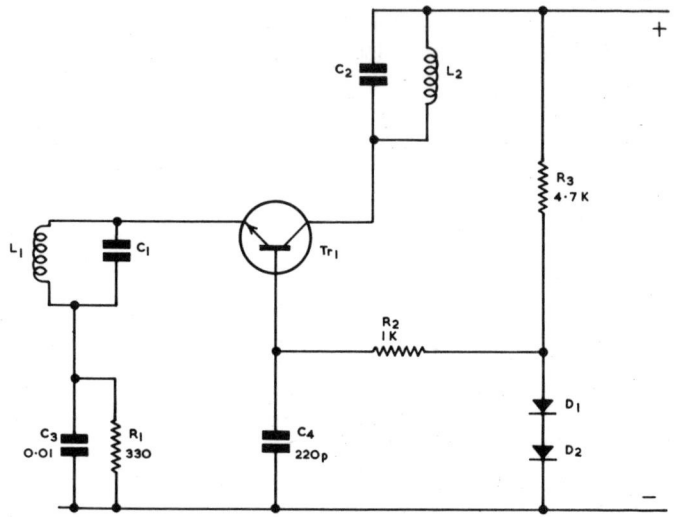

FIG. 2.19. BIAS ARRANGEMENTS OF R.F. AMPLIFIER OR MIXER

R_1, which is by-passed by C_3. The collector is taken to the positive line through the output tuned circuit $L_2\ C_2$. In this case a low base voltage is required equal to the drop across R_1 (1 volt or less) and the base-emitter drop of the transistor. The base voltage is obtained from the drop across the two forward-biased diodes D_1 and D_2, which are fed from the positive line through R_3. C_4 by-passes the base to earthy line, and R_2 together with C_4 is decoupling. A number of transistors may be fed from across $D_1\ D_2$. This arrangement has the advantage that the standing transistor collector current will be more constant than when the base is fed from a potential divider, both as regards variations of temperature and supply voltage. If the temperature increases, the collector current rises for a given base-emitter voltage (V_{be}). But the voltage across D_1 and D_2 will fall so reducing the bias voltage and tending to maintain a constant collector current. If the supply voltage varies (as it will if the set is battery operated) the voltage across the diode will remain approximately constant (due to their characteristics) and hence the collector current will remain approximately constant. The same basic circuit can be used for the "upside-down" circuit, e.g. figure 2.3, but the diodes are then replaced by a zener diode.

If the supply voltage to the oscillator varies it may cause variations in the oscillator frequency and can be reduced by feeding the oscillator through a

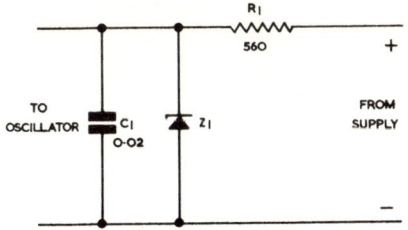

FIG. 2.20. VOLTAGE STABILIZER CIRCUIT FOR OSCILLATOR

simple voltage stabilizer circuit consisting of a resistor and zener diode. Such a circuit is shown in figure 2.20. Variations in supply voltage will vary the current in R_1, and therefore the current in the zener diode Z_1. However, the voltage across a zener diode is almost independent of the current flowing through it: hence a constant voltage is fed to the oscillator. C_1 is a by-pass for any h.f. current. In some circuits a diode may be placed across the input circuit of the r.f. amplifier, across its output circuit, or across the i.f. transformer following the oscillator. The diode acts as limiter and prevents an excessive voltage being fed into later stages.

I.F. AMPLIFIERS

The i.f. amplifier is required:

(a) To increase the magnitude of the i.f. voltage from the mixer to a sufficiently high value to operate the demodulator.

(b) To provide the main selectivity of the receiver so as to remove adjacent stations. The frequency response should be uniform over, say, 250 kHz so as to amplify all the sidebands of a given station by the same amount, but outside this the response should fall sharply.

(c) To act probably as a limiter so that the voltage fed to the demodulator is of constant value.

The intermediate frequency universally used is 10·7 MHz.

Considering first the small receiver, the i.f. amplifier normally uses double-tuned transformers between transistors. These transformers act as bandpass circuits, and give the type of frequency response which is desirable. Usually three stages are used, i.e. three transistors and four transformers. Since the frequency is now relatively low common-emitter circuits are usually used. A circuit is shown in figure 2.21.

In this circuit Tr_1 is fed through C_1 from the mixer, but a transformer would be more usual. A transformer is also required between Tr_3 and the demodulator. Bias for Tr_1 is provided by R_1 and R_2, R_3 being the emitter stabilizing resistor. The emitter is by-passed to the earthy line by C_2 (as it must be if it is operating as a common-emitter stage). The first transformer is T_1 with both primary and secondary tuned. A tap is used to reduce the damping on the secondary, by the low input impedance of Tr_2. The other transistors operate in a similar manner. C_3 and C_5 complete the i.f. circuit from the lower end of the transformers to the earthy line. If they become open circuit the gain is much reduced as the i.f. current must then flow in R_5 or R_8. The bandwidth of the amplifier depends on the Q of the tuned circuits in the transformers. In place of double-tuned transformers only the primary may be tuned. Sometimes double transformers are used, i.e. two transformers coupled together; both having two tuned circuits. Due to the collector-base capacitance a current is fed through this capacitance and it may cause instability or oscillation. This can be overcome by what is called

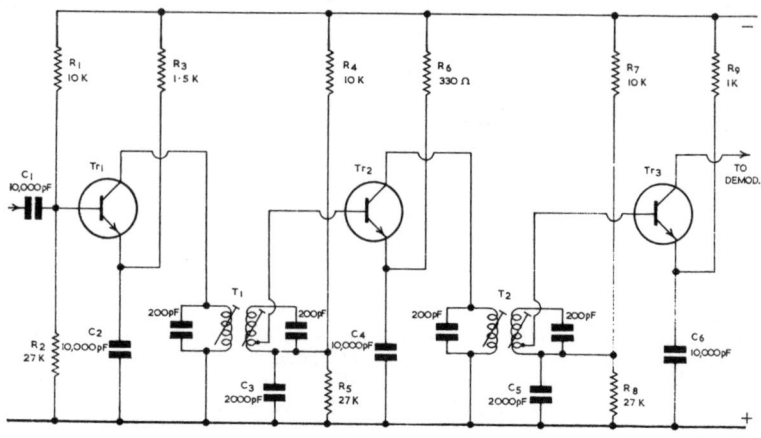

FIG. 2.21. I.F. AMPLIFIER

neutralizing. The idea is to feed a current to the base, from the collector, through another circuit and arrange that this current is equal and 180° out of phase with that through the collector-base capacitance. This may be done in two ways, as shown in the basic circuits of figure 2.22.

L_1C_1 form the primary of the transformer, and L_2C_2 the secondary. At (a) the primary is tapped. For simplicity assume a centre tap. In this case the voltages at the top and bottom of L_1, relative to the positive line, will be equal and of opposite phase (*i.e.* 180° out of phase). A feedback current will flow through C_{cb} (the collector-base capacitance) and by connecting C_3 of value equal to C_{cb}, an equal current will flow in C_3 and will cancel that in C_{cb} since it is 180° out of phase. A centre tap is not essential in which case C_3 will not now be equal to C_{cb}. As C_{cb} may not be a perfect capacitor, a resistor can be placed in series with C_3 to ensure that the currents are in antiphase. At (b) the operation is similar, but the voltage for feeding a current through C_3 is taken from the secondary. The secondary must, of course, be so wound that the voltage fed to C_3 is 180° out of phase with that at the collector, but there is no difficulty in doing this. Neutralizing is not commonly used because the capacitance of modern transistors is small enough to prevent oscillation.

In place of the electrical transformer a ceramic filter or filters may be used. This makes use of the piezo-electric property of certain ceramic materials. A material which is piezo-electric has the property that, when an electric field is applied to it, say by the use of two conducting coatings deposited on it, it will change its dimensions. Similarly, if the material is subject to a mechanical force it will generate an e.m.f. between the two plates. If an alternating voltage is applied to such a device it is found that it behaves in a similar way to an electrical resonant circuit. This is due to the fact that the alternating voltages cause it to vibrate and mechanical resonance occurs at some frequency, this frequency being determined by the physical size of the device and on the method of vibration. The Q of such a device is much higher than can be obtained in an electrical circuit, and hence it forms a very good filter, giving good selectivity. It may be made to act as a transformer by using two coatings such as shown in figure 2.33 on one side and a continuous coating on the other side, the coating being connected to the earthy line. The input is fed to one coated area and the output taken from the other. By suitable design the impedance of the device can be made to suit the transistors. Since the ceramic is an insulator suitable

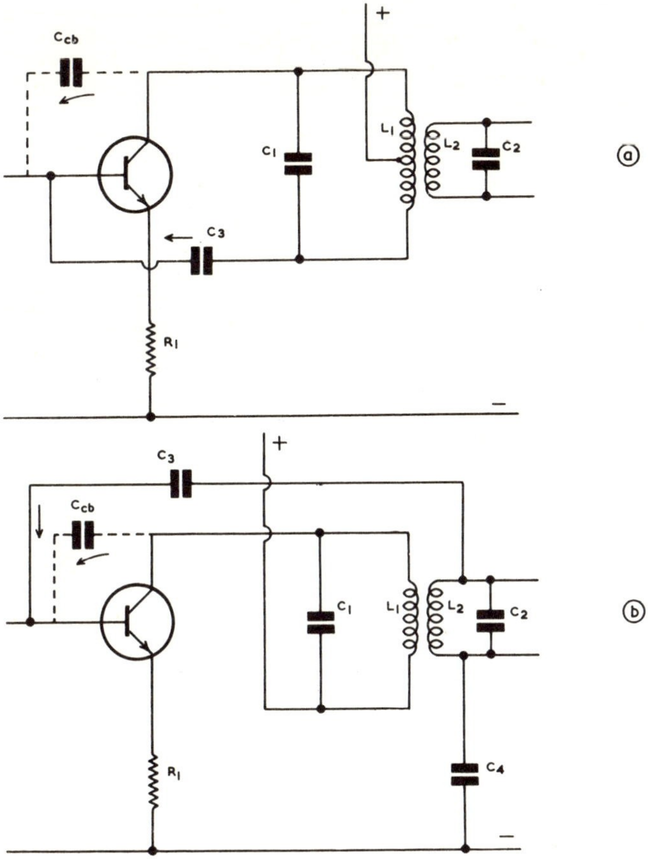

FIG. 2.22. NEUTRALIZING CIRCUIT, (a) USING TAPPED INDUCTOR; and (b) USING FEEDBACK FROM THE SECONDARY

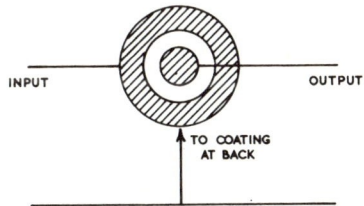

FIG. 2.23. CERAMIC FILTER

direct voltages must be fed to the transistors, an example being given in figure 2.24. R_1 is the d.c. feed to Tr_1, and R_3R_4 provide bias for Tr_2. Ceramic filters are more commonly used on tuner units than small portable receivers, and several such filters may be used in a good tuner unit. The frequency of the filter is, of course, fixed and nominally 10·7 MHz. However, there will be some tolerance, and the frequency of maximum response may not be exactly 10·7 MHz. If not, then any alignment must allow for this and other i.f. components adjusted to the frequency of the ceramic filter. When more than

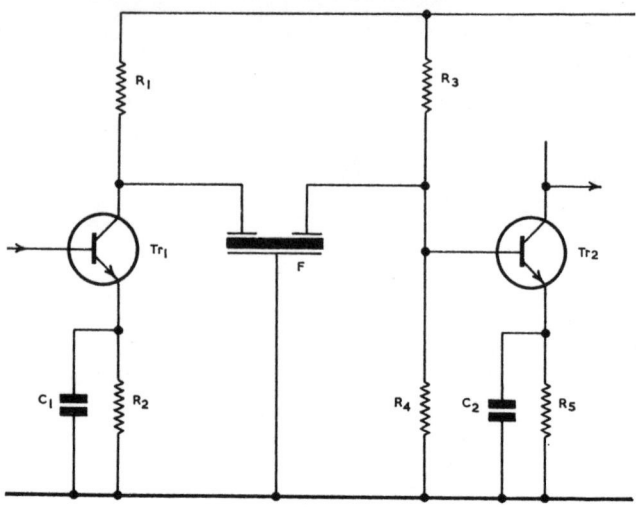

FIG. 2.24. USE OF CERAMIC FILTER IN I.F. AMPLIFIER

one filter is used they must be of the same frequency; they are often colour coded or only supplied as a set of filters.

When a single filter is used the other stages of the i.f. amplifier may use transformers in the normal manner, or an untuned circuit may be used as in figure 2.25. Transistor Tr_1 has a load R_1 which will be fairly low to reduce the effect of stray capacitance across it. The a.c. component is fed through C_1 to Tr_2, and R_3 and R_4 provide bias for Tr_2.

Integrated circuits are now being used in i.f. amplifiers. One problem when using an integrated circuit is that it cannot itself be made selective since, obviously, tuned circuits cannot be incorporated. One solution is to provide the selectivity before the i.c. by using, for example, one or more ceramic filters. The i.c. is then a flat amplifier. Alternatively, tuned circuits

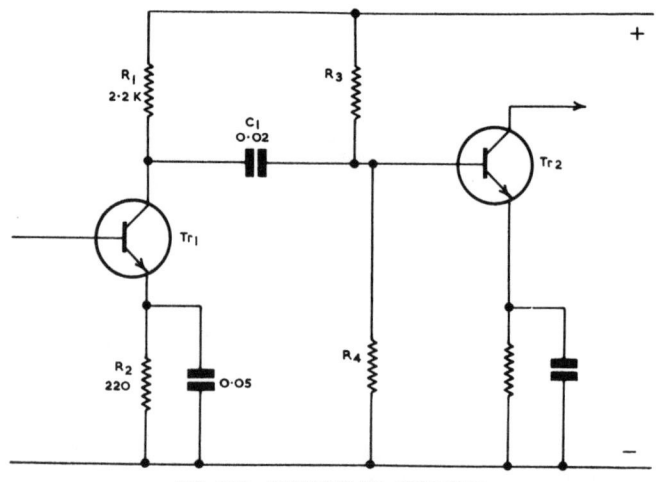

FIG. 2.25. UNTUNED I.F. AMPLIFIER

may be used external to the i.c. In this case there may be one stage of flat amplification followed by a selective stage with an external i.f. transformer followed by another amplification stage feeding the demodulation through a transformer. I.C.s used for i.f. amplification often include other sections of the receiver, such as the demodulator. Accordingly, there are many possibilities some of which will be considered in the section on demodulators.

In some cases it is desirable for the i.f. amplifier to act as a limiter, so that any a.m. is removed before the demodulator. Limiting will occur in any i.f. amplifier if the signal is large enough because, in one direction, the base-emitter junction will be reverse biased, and in the other direction the transistor will be bottomed (*i.e.* the collector current will be large enough to reduce the collector voltage to almost zero). However, diodes may be used for limiting, and an example is given in figure 2.26. Back-to-back diodes D_1 and D_2 are used so that symmetrical clipping or limiting occurs. R_1 and the diodes form

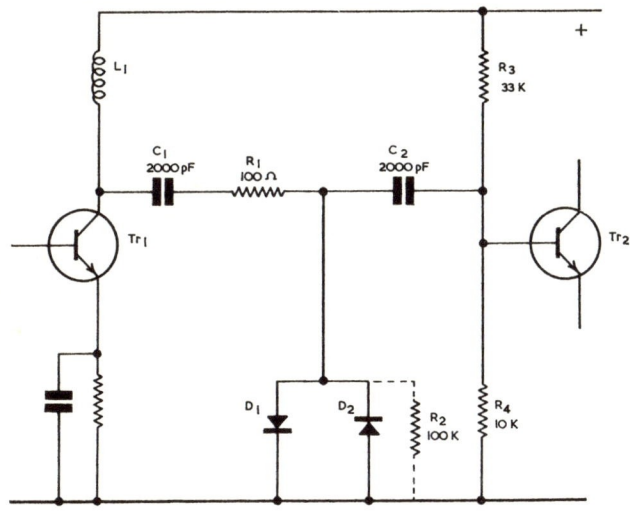

FIG. 2.26. LIMITER CIRCUIT

a potential divider. R_1 has little effect until the diodes conduct, but once they do the output from across the diodes is limited to an almost constant voltage (say $\frac{1}{2}$ volt). C_1 and C_2 are blocking capacitors to prevent d.c. being applied to the diodes, and R_3R_4 provides bias for Tr_2. R_2 may be added to prevent a steady voltage being developed across D_1 and D_2, which could occur if the characteristics were not similar.

F.M. DEMODULATORS

Until recently only two types of demodulator were commonly used: the Foster-Seeley circuit and the ratio detector. The Foster-Seeley circuit will not be described because it is rarely used in domestic receivers in this country. The circuit looks like that of the ratio detector, but the method of operation is considerably different. The ratio detector is used in practically all receivers apart from those using demodulator integrated circuits. The type of demodulator in an integrated circuit is quite different and will be described later.

The basic balanced ratio detector circuit is shown in figure 2.27. In the collector circuit of the transistor is the tuned circuit L_1C_1, which is tuned to the centre frequency, *i.e.* 10·7 MHz. This circuit is coupled by mutual inductance to the secondary L_2C_2, which is also tuned to 10·7 MHz. L_3 is a coupling

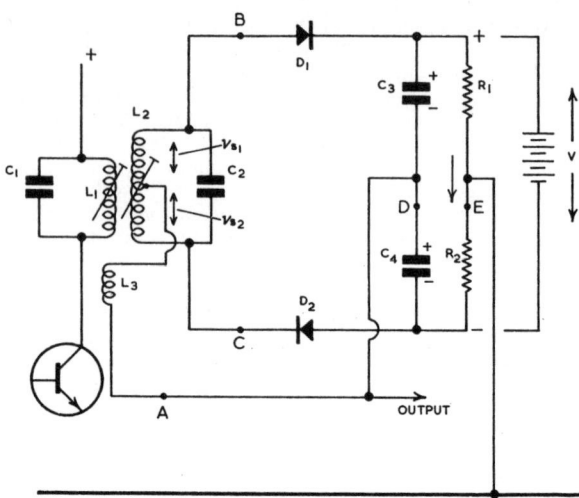

FIG. 2.27. BASIC RATIO DETECTOR CIRCUIT

winding, coupled by mutual inductance, to the primary L_1. The circuit uses the fact that, at the centre frequency, the voltages in the secondary v_{s1} and v_{s2} are at 90° to the voltage induced in L_3. As the frequency deviates from this value the angle changes, the voltages no longer being at 90°. To show this we must consider the phasor diagram of figure 2.28, at the centre frequency.

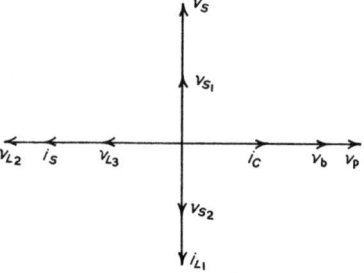

FIG. 2.28. PHASOR DIAGRAM OF RATIO DETECTOR AT THE CENTRE FREQUENCY

Starting with the base voltage v_b, this will cause a collector current i_c to flow, which will be in phase with v_b. Since the primary circuit is at resonance, the voltage v_p across the primary will be in phase with the collector current i_c. The current i_{L1} which flows in the primary winding L_1 will lag this voltage by 90°—assuming a perfect inductor. This is shown in figure 2.28. The current flowing in L_1 will induce a voltage v_{L2} in the secondary winding L_2 and a voltage v_{L3} in winding L_3, and this will lag the current i_{L1} by 90°. The reason is that the voltage across an inductor leads the current by 90° (or the current lags the applied voltage by 90°), but the e.m.f. induced in the inductor is 180° out of phase with the applied voltage, so that it opposes it. Thus the induced voltage is 90° lagging the current. The voltage v_{L2} induced in L_2 will cause a current i_s to flow in the secondary circuit L_2C_2, which acts as a series circuit as regards this induced e.m.f. Since this circuit is also at resonance, it behaves as a resistance and hence the current i_s is in phase with the induced e.m.f. v_{L2}. This current flows through C_2, hence the voltage v_s across C_2 (and hence across L_2), will lag the current by 90° (the current in a capacitor leads the voltage by 90° or the voltage lags the current by 90°). The secondary voltage v_S can be divided into two voltages v_{S1} and v_{S2}, which, as regards the centre tap, are 180° out of phase, as shown in the figure. We will now see what currents flow as a result of these voltages. The voltage between A and B is the vector addition of v_{L3} and v_{S1} and shown as v_{AB} in figure 2.29. In a similar way the voltage between A and C (v_{AC}) is the vector sum of v_{L3} and v_{S2} and this is shown in the same figure. v_{AB} and v_{AC} are numeri-

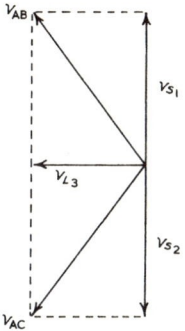

FIG. 2.29. VOLTAGES IN RATIO DETECTOR CIRCUIT AT THE CENTRE FREQUENCY

cally equal. Now, the voltage v_{AB} will cause a current to flow through D_1 and charge up C_3 to the peak value of v_{AB}, in the direction shown. v_{AC} will also cause a current to flow through D_2 and charge up C_4 in the direction shown. Since v_{AB} and v_{AC} are numerically equal, then V_{C3} must equal V_{C4}. Resistor R_1 is equal to R_2, therefore there must be equal voltages across R_1 and R_2, and these voltages must equal those across C_3 and C_4. Thus R_1, R_2, C_3 and C_4 form a balanced bridge circuit and there is no voltage between D and E.

We will now consider what happens if the i.f. is higher than its centre value of 10·7 MHz. The phasor diagram is drawn in figure 2.30. Starting off with the base voltage the collector current will be approximately in phase

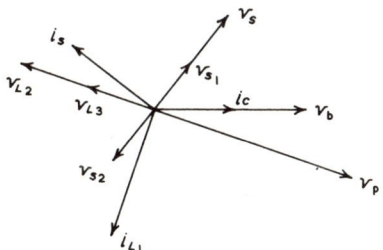

FIG. 2.30. PHASOR DIAGRAM OF RATIO DETECTOR ABOVE THE CENTRE FREQUENCY

with it. Unlike the last case, the primary circuit is now not so resonant and a larger current will flow in C_1 than L_1 (because the frequency is above the resonant frequency), therefore the circuit behaves as a capacitance. The voltage across the circuit will lag the current (*i.e.* the current will lead the voltage). This voltage v_p will cause a current i_{L1} to flow in L_1, which lags it by 90°, and will induce a voltage v_{L2} and v_{L3} 90° lagging this current as previously explained. However, the secondary circuit is not at resonance, and because the frequency is above the resonant value the inductive reactance will be greater than the capacitive reactance (the circuit being series as regards the induced e.m.f. v_{L2}). Thus the current i_S will lag the induced e.m.f. The voltage across the capacitor C_2 (v_S) will lag the current by 90° as shown. It is now seen that v_{S1} and v_{S2} are no longer at 90° to the voltage induced in L_3, v_{L3}. Figure 2.31 shows part of this phasor diagram with V_{L3} horizontal corresponding to figure 2.29. It will now be seen that v_{AB} is no longer equal

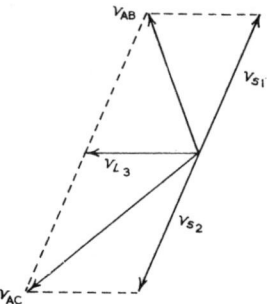

FIG. 2.31. VOLTAGE IN RATIO DETECTOR CIRCUIT ABOVE THE CENTRE FREQUENCY

to v_{AC} in magnitude, the latter being greater. We will now see the effect on the circuit diagram. Since v_{AB} is now less the voltage across C_3 will be reduced, and since v_{AC} is greater the voltage across C_4 will be increased. The sum of the two voltages is approximately constant, and the voltages across R_1 and R_2 (which must be equal, since $R_1 = R_2$) will be approximately the same as previously. Hence, point D is more positive than point E, and since E is connected to the earthy line the output is positive. The greater the deviation in frequency the greater will be the change of phase angle from 90° and the greater the positive output (within limits). If the frequency change is in the opposite direction, i.e. a decrease of frequency, then the phase change will be in the opposite direction and v_{AB} will be greater than v_{AC}. As a result there will be a negative voltage output. Thus, as the frequency swings back and forth the output will change approximately in proportion to the frequency change. A typical relationship between frequency and output is shown in figure 2.32. For no distortion the voltage output should be proportional to the change in frequency, and this is seen to be approximately true over the

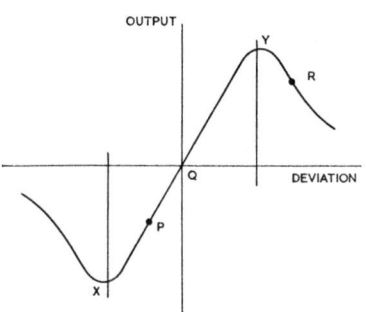

FIG. 2.32. CHARACTERISTICS OF RATIO DETECTOR

range X to Y. These two points must, of course, be far enough apart on the frequency scale to cover the maximum deviation. Since the deviation we are concerned with is ± 75 kHz, points X and Y must be at least 150 kHz apart and preferably, say, 200 kHz. For no distortion the receiver must be accurately tuned, so that the unmodulated carrier corresponds to point Q. If tuning is not accurate then the demodulator might operate from P to R and serious distortion will then occur. We will return to the need for accurate tuning later in the chapter.

As drawn the circuit in figure 2.27 will respond to amplitude modulation. If the input increases then the voltages across C_3, C_4, R_1 and R_2 will all increase, as will the output. Thus, any variations in amplitude due to amplitude modulation will appear at the output (except at the centre frequency, when the output is zero). As we have seen, when the voltage V_{C3} changes in one direction the voltage V_{C4} changes in the opposite direction, and so the total voltage does not vary appreciably with deviation or frequency modulation. Thus we could place a battery across R_1 and R_2, as indicated in figure 2.27, the battery voltage being equal to that normally across these resistors. Any amplitude modulation would now not produce any output because the battery would prevent variations in the total voltage across R_1 and R_2. A battery is not convenient and it can be replaced by a LARGE capacitor C_6, as in figure 2.33. This capacitor will become charged to the mean voltage across R_1 and R_2, and any rapid variations due to amplitude modulation (mainly by noise)

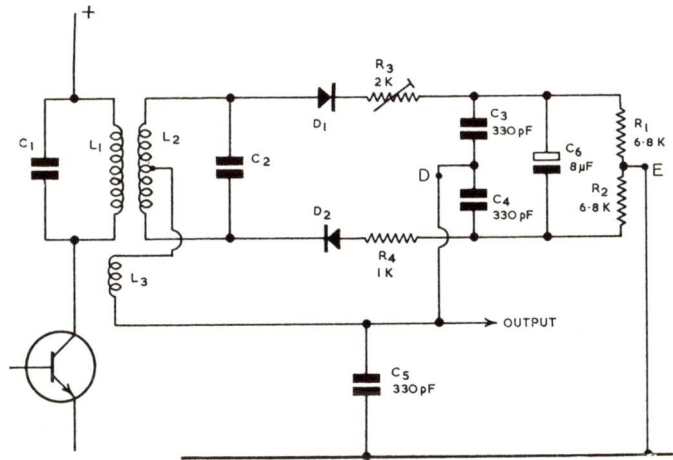

FIG. 2.33. RATIO DETECTOR CIRCUIT

will be shorted out by this capacitor. If, of course, the mean value of the signal to the demodulator changes, then C_6 will change its voltage to the appropriate value. If there is to be no output due to a.m. the circuit should be balanced, but there will almost certainly be some difference in the diode characteristics. This is reduced by placing a fixed resistor R_4 in series with one diode, and a preset resistor R_3 in series with the other, R_3 being adjusted for minimum output with amplitude modulation. C_5 completes the circuit to the earthy line as regards the i.f. Instead of connecting point E to the earthy line, point D may be so connected and the output taken from point E. The operation is identical.

This is a balanced circuit, and a number of unbalanced circuits are possible. One such circuit is given in figure 2.34. In this case there will be the normal audio frequency output, but also the steady voltage across R_2. Provided this d.c. component is not fed to the a.f. amplifier it has no effect on the operation of the demodulator and a.f. amplifier. Instead of using a tertiary winding L_3 a voltage proportional to the primary voltage can be obtained from a capacitor potential divider, as in figure 2.35. The voltage across C_7 is now

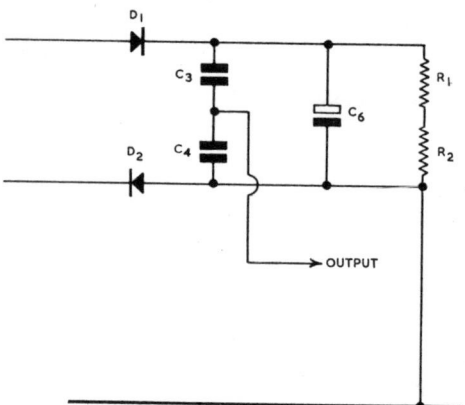

FIG. 2.34. UNBALANCED RATIO DETECTOR CIRCUIT

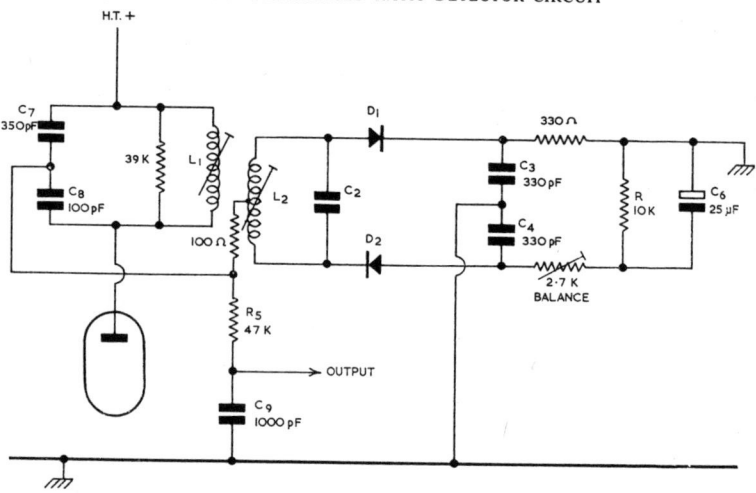

FIG. 2.35. RATIO DETECTOR CIRCUIT NOT USING TERTIARY WINDING

applied across R_5 and is in series with L_2 as previously. Although this, and
the next circuit, are shown as unbalanced circuits, they could have been shown
as balanced circuits. Another modification is shown in figure 2.36, in which
the voltage corresponding to that in L_3 is obtained from across C_2, C_1 and
C_2 forming a potential divider across the primary L_1. Instead of using a
tapped inductor for L_2, the centre tap is obtained by using two equal capacitors
C_3 and C_4.

Pre-emphasis and de-emphasis were considered in Chapter 1. At the
receiver de-emphasis must be used to restore the signal back to the original,
which is done by passing the output of the demodulator through a de-emphasis
circuit, consisting of a series R and parallel C of time-constant 50 μs. This is
shown in figure 2.37. In this case the time constant (*i.e.* C $\times$ R) is 47×10^3
$\times 1000 \times 10^{-6} = 47$ μs. The 0·047 μF capacitor is a d.c. blocking capacitor
to prevent any direct voltage being applied to the a.f. amplifier.

Integrated circuits are now being used for the demodulator of f.m. receivers,
but an i.c. does not lend itself to either the Foster-Seeley demodulator or ratio

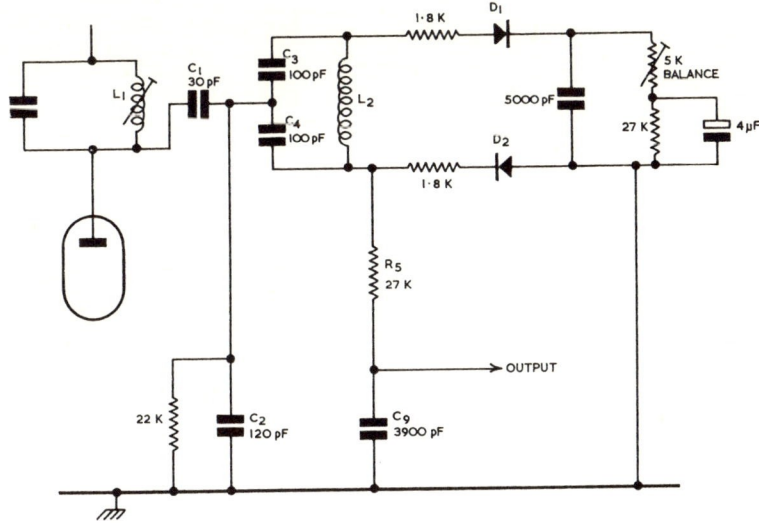

FIG. 2.36. RATIO DETECTOR CIRCUIT NOT USING A CENTRE-TAPPED INDUCTOR

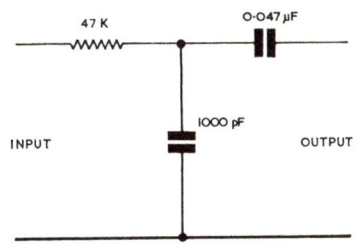

FIG. 2.37. DE-EMPHASIS CIRCUIT

detector circuits, and a new principle is used. It is hardly worthwhile just using an integrated circuit for the demodulator, so the integrated circuit performs other functions as well, in particular acting as an i.f. amplifier and limiter. The i.f. amplifier and limiter will be described first. This generally makes use of long-tailed pair stages, and a basic i.c. circuit is given in figure 2.38. The two transistors Tr_1 and Tr_2 are connected with a common emitter resistor R_3, this arrangement being commonly known as a **long-tailed pair**. In some i.c.s R_3 and R_6 are replaced by constant-current circuits which give improved performance. It is shown with two inputs, 1 and 2, which are in antiphase (180° out of phase). but only one input is necessary. Suppose that we have a signal on input 1, and that input 2 is connected to a suitable bias supply. Normally, both transistors will be conducting. If the input now increases in a positive direction the current of Tr_1 will increase, which will increase the current in, and the voltage across, R_3. However, as the voltage increases across R_3 it reduces the base-emitter voltage of Tr_2, and so reduces its collector current. If both transistors are conducting, the current in R_3 will remain approximately constant. However, when the input to Tr_1 exceeds a certain value the voltage across R_3 will rise to such a value that Tr_2 is cut off, and no collector current will flow. Hence Tr_2 limits positive inputs above a certain value. If the input to Tr_1 now goes in a negative direction, the collector current of Tr_1 decreases, so decreasing the voltage across R_3, which increases

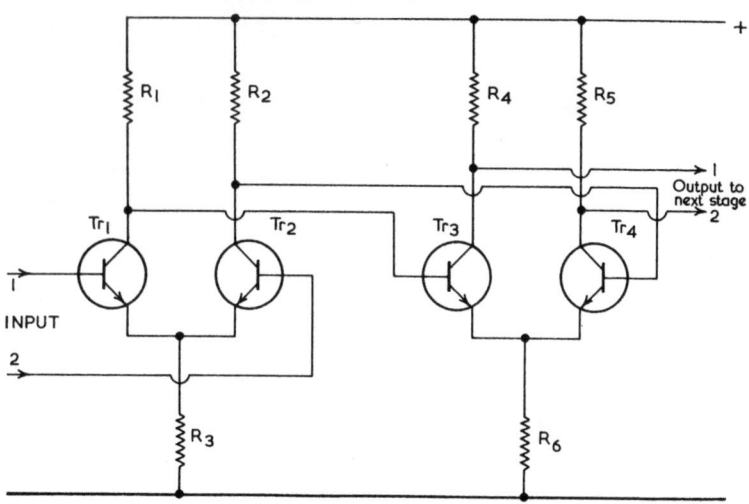

FIG. 2.38. INTEGRATED CIRCUIT I.F. LIMITING AMPLIFIER

both the base-emitter voltage of Tr_2 and its collector current. Eventually the current of Tr_1 is reduced to zero and Tr_1 is cut off, limiting now occurring in the negative direction. This circuit acts therefore as a very effective limiter, the voltage range over which limiting occurs being quite small. When both inputs are fed with signals in antiphase, the operation is similar, except that when the current of one transistor is increased by the signal, the other is decreased, not only by feed through the common emitter resistor but also by the signal on its base. To increase the amplification and limiting, a number of stages may be used in cascade, two such stages being shown in the figure 2.38. An alternative arrangement is given in figure 2.39. In this case emitter-followers are used between stages of long-tailed pair transistors, and in one i.c. six stages of long-tailed pairs are used. This results in very good limiting with a.m. suppression of greater than 50 dB. The input required for limiting is small (equivalent to a large gain) and may be as low as 100 μV.

Since this i.c. i.f. amplifier gives no selectivity this must be provided before the i.c. (there are no arrangements for inserting tuned circuits in the amplifier or between the amplifier and demodulator). This selectivity may be provided by suitable electrical tuned circuits or by ceramic filters.

Turning now to the demodulator section, a simplified circuit is shown in figure 2.40. This type is known as a **quadrature demodulator.** Transistors Tr_1 and Tr_2 form a long-tailed pair with R_1 as the common-emitter resistor which, in practice, is usually replaced by a constant-current circuit. The bases of Tr_1 and Tr_2 are fed in antiphase from the limiting amplifier so that these are of square waveform. Antiphase inputs are not essential: only one base need be fed. A tuned circuit L_1C_1 is connected across the antiphase inputs (through small capacitors C_2 and C_3) having appreciable reactance. The tuned circuit L_1C_1 is tuned to 10·7 MHz and therefore, at the centre frequency, it behaves as a resistance. Hence, the voltage across this circuit will be approximately 90° out of phase with the voltages applied to the bases of Tr_1 and Tr_2. The voltage across this tuned circuit is used to switch transistors Tr_3, Tr_4, Tr_5 and Tr_6. Tr_3 and Tr_5 have a common load R_2; Tr_4 and Tr_6 have a common load R_3. R_2 can be omitted since the voltage across it is not used. The operation of this circuit is best explained by the waveform diagram of figure 2.41. At (a) are shown the conditions when the frequency is at its centre value of 10·7

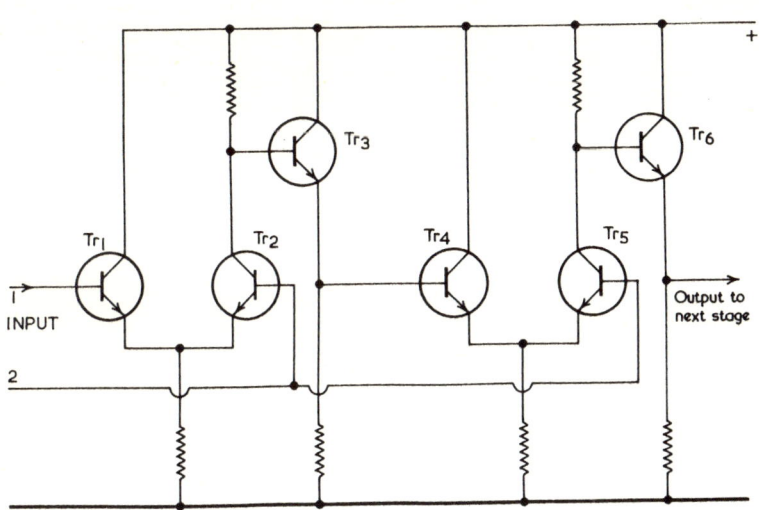

FIG. 2.39.   ALTERNATIVE INTEGRATED CIRCUIT I.F. LIMITING CIRCUIT

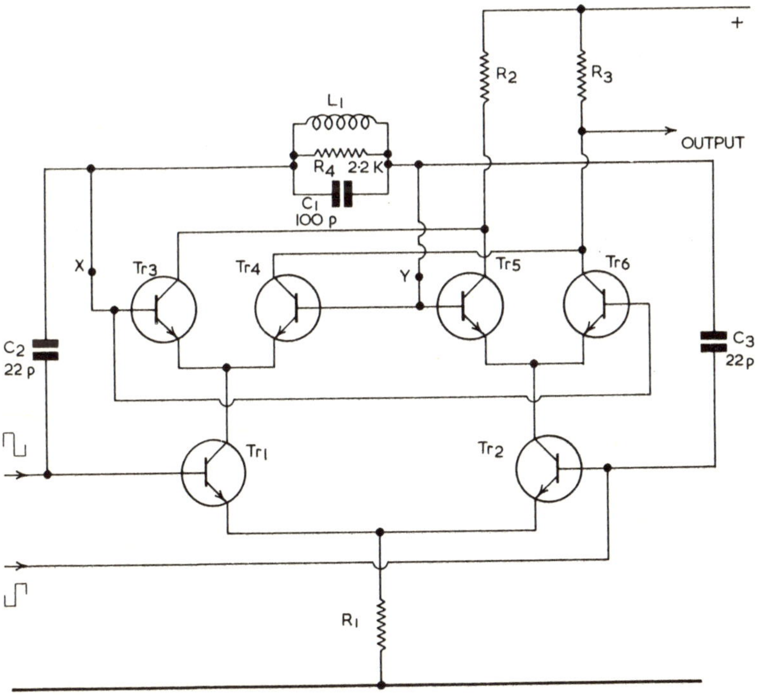

FIG. 2.40.   INTEGRATED CIRCUIT QUADRATURE DEMODULATOR

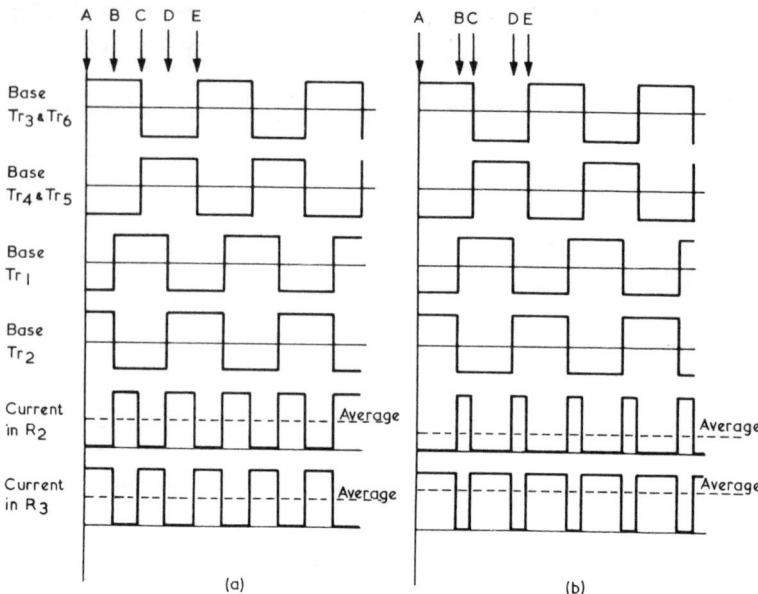

FIG. 2.41. OPERATION OF QUADRATURE DEMODULATOR, (a) AT CENTRE FREQUENCY; and (b) WITH DEVIATION FROM THE CENTRE FREQUENCY

MHz. Perfect square waves will be assumed for simplicity. The bases of Tr_3 and Tr_6 are fed from one side of the tuned circuit, and are fed with signals 180° out of phase with those on the bases of Tr_4 and Tr_5, which are fed from the other side of the tuned circuit. The voltage from the tuned circuit is sufficiently large to switch the transistors from the fully ON to the fully OFF condition. These voltages are also 90° out of phase with the voltages on the bases of Tr_1 and Tr_2, which, since they are in the emitter circuits of the other transistors, control the current flow in them. Current, for example, will only flow in R_2 if Tr_3 AND Tr_1 are ON or Tr_5 AND Tr_2 are ON. Between instants A and B Tr_3 and Tr_6 are ON and Tr_2 is ON, and a current will flow in Tr_2 and Tr_6 and R_3, as shown in the figure. No current can flow in R_2 since in one circuit Tr_5 is OFF and in the other possible circuit Tr_1 is OFF. From B to C Tr_3 and Tr_6 are still ON, but now Tr_1 is ON, instead of Tr_2. Current now flows in Tr_1 and Tr_3 through R_2. No current can flow in R_3 because in one circuit Tr_2 is OFF and Tr_4 is OFF in the other circuit. In the same way from C to D current flows in R_3, and from D to E current flows in R_2. Thus, the currents in R_2 and R_3 are equal in value and consist of rectangular waves having equal ON to OFF ratios, the mean value being half the peak value.

Suppose now that the incoming frequency is deviated so that it is no longer 10·7 MHz. The tuned circuit will no longer behave as a resistor, and the voltage across it will no longer be 90° out of phase with the input voltage. Suppose that it is at an angle of 45° to the input voltage, as shown in figure 2.41(b). From A to B Tr_3 and Tr_6 are ON, as also is Tr_2. Thus there is now a circuit through Tr_2, and Tr_6 and R_3. There is no other conducting circuit so no current flows in R_2. From B to C Tr_3 and Tr_6 are still ON, but now Tr_1 is ON. Thus, the circuit for current flow is Tr_1, Tr_3 and R_2. From C to D Tr_4 and Tr_5 become conducting together with Tr_1. Now current flows only in Tr_1, Tr_4 and through R_3. From D to E Tr_4 and Tr_5 are still ON, but Tr_2 now is made conducting. The circuit for current flow is now Tr_2, Tr_5 and R_2. It is now seen that equal currents do not flow in R_2 and R_3, the mean value of that in R_2 being less than that in R_3. What is more important is that the mean currents are different from those shown at (a). Thus, for frequency deviation in this particular direction the current in R_2 decreases, while that in R_3 increases. If the frequency deviation had been in the other direction then it is fairly easily shown, by drawing appropriate waveform diagrams, that the current in R_2 would be greater than that at (a), while that in R_3 would be less. Thus, as the frequency deviates back and forth, the currents in the two resistors vary accordingly. If the circuit is designed correctly there is almost a linear relationship between the output current (in R_3), and hence output voltage, and frequency over a limited range. A typical characteristic is shown in figure 2.42. The frequency difference between points X and Y must be sufficient to cover

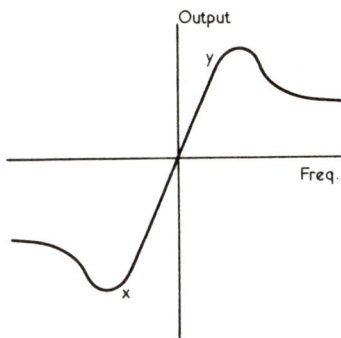

FIG. 2.42. RESPONSE CURVE OF QUADRATURE DEMODULATOR

the maximum deviation of the system, in our case ± 75 kHz. This range is determined by the damping or Q of the tuned circuit $L_1\,C_1$. The advantage of this type of demodulator is that there is no fundamental i.f. output. This helps in maintaining stability and only a single tuned circuit is required. Alignment is therefore easy. A capacitor, external to the i.c., is used to remove the second harmonic of the i.f. and, if of the correct value, produces the required de-emphasis. Since the demodulator is a switching circuit the output depends only on the frequency deviation and not on the magnitude of the input (provided it is large enough to switch properly), and so the circuit automatically suppresses any a.m. and may, in fact, operate in this limiting or switching way at inputs where the i.f. amplifier may not limit. Normally the input to the i.f. amplifier is such that limiting occurs both in the i.f. amplifier and in the demodulator, giving excellent a.m. rejection.

Where antiphase signals are not available then Tr_1 is fed with the input, but Tr_2 will also switch due to the feed through the common emitter resistor R_1. In this case the tuned circuit is fed through a single small capacitor and the other end connected to the earthy line. Tr_3 and Tr_6 are fed from the live side of the tuned circuit, and since Tr_3 and Tr_4 form a long-tailed pair, if Tr_3 is switched on its emitter voltage will rise and so cut off Tr_4. Similarly, if Tr_3 is cut off the emitter voltage will drop, causing Tr_4 to conduct. A similar action takes place with Tr_5 and Tr_6. When only Tr_1 is fed with a signal Tr_2 must be fed with a suitable bias voltage (*i.e.* the same as the steady voltage on Tr_1). The actual integrated circuit is more complex than that shown. Constant current circuits may be used in place of the common-emitter resistors, and emitter-followers are used to reduce the loading between circuits. Such a type of demodulator could, of course, be built with discrete components, but the cost would be excessive compared with that of a ratio detector.

A.F. AMPLIFIER

After the demodulator the a.f. signal is fed to the a.f. amplifier and output stages. Small power circuits have been described in *Radio Servicing*, Volume 2, and such circuits are used in f.m. receivers. Integrated circuits may be used. In the case of tuner units, these feed high-power hi-fi amplifiers and are described in Chapter 3.

AUTOMATIC GAIN CONTROL (A.G.C.)

As a limiting amplifier is commonly used in an f.m. receiver one might expect that a.g.c. would not be used. In fact, since limiter circuits operate better with a large input, a.g.c. would appear to be a disadvantage. However, there are cases where cross-modulation can occur, *e.g.* in the mixer when there is a large signal input. It is desirable, therefore, to reduce the gain of the r.f. stage on large signals. Note that the a.g.c. voltage is applied only to the r.f. stage. Since the input to the demodulator tends to be constant (due to use of limiter circuits or the automatic limiting which occurs on a normal i.f. amplifier), the a.g.c. control voltage cannot usually be obtained from the demodulator. It must be obtained from an early stage in the i.f. amplifier before limiting occurs.

A simple circuit is shown in figure 2.43. This shows a typical r.f. amplifier stage with $L_1 C_1$ the input circuit and $L_2 C_4$ the output stage. The collector is

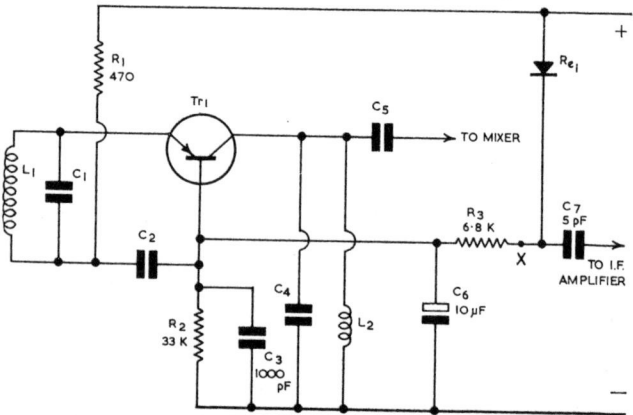

FIG. 2.43. AUTOMATIC GAIN CONTROL CIRCUIT FOR F.M. RECEIVER

returned to the earthy supply line through L_2 and the emitter is taken to the positive line through L_1 and the stabilizing resistor R_1. Bias for the base is provided by the potential divider $R_2 R_3$, which is fed through the forward conducting diode Re_1. Capacitor C_7 is fed from the first stage of the i.f. amplifier (before limiting takes place). When there is a large i.f. signal this is rectified by Re_1 so that point X goes more positive. This increases the voltage on the base and therefore reduces the base-emitter voltage on Tr_1. This will reduce the collector current and the gain of the transistor, so reducing the voltage fed to the mixer and reducing the possibility of cross-modulation due to overloading. C_6 is added to remove any i.f. voltage or fluctuations due to amplitude modulation of the i.f. signal.

Another circuit is given in figure 2.44, an r.f. stage using an F.E.T. Here a voltage-doubler rectifier, consisting of D_1, D_2, C_4 and C_5, is used to generate the control signal. This voltage-doubler is fed from the first stage of the i.f. amplifier. When the input is positive D_1 conducts and charges up the capacitor C_5, as shown. When the input is negative the voltage of C_5 is added to the

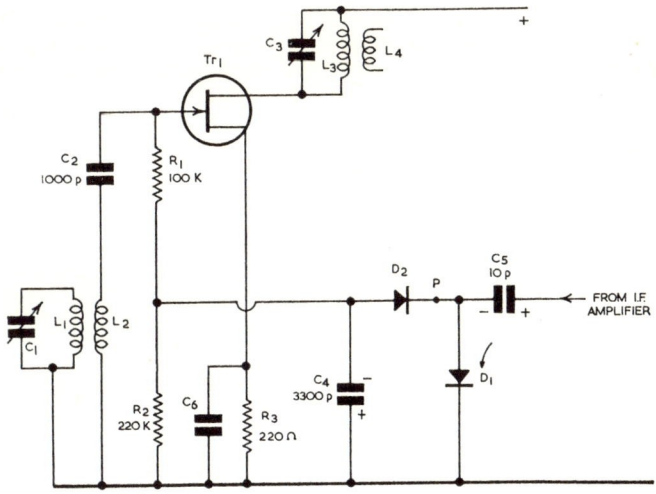

FIG. 2.44 AUTOMATIC GAIN CONTROL CIRCUIT APPLIED TO AN F.E.T. AMPLIFIER

input and charges C_4 through D_2. Thus the voltage across C_4 approaches twice the peak input voltage. The negative voltage on C_4 is fed as bias to the gate of Tr_1, so reducing its gain. R_1 must be added to prevent C_4 shorting out the signal input on the gate, and R_2 allows C_4 to discharge at a reasonable rate when the magnitude of the i.f. signal decreases.

TUNING INDICATORS AND AUTOMATIC FREQUENCY CONTROL (A.F.C.)

It was pointed out when dealing with demodulators that accurate tuning was essential if serious distortion was to be avoided. Accurate tuning of an f.m. receiver is not easy. When the percentage modulation is small (as it is most of the time) the operating point is not very critical, so long as the demodulator operates on the straight part of the characteristic. Since this straight part is longer than the deviation at low percentage modulations, errors in tuning make little difference. However, when a large sound occurs, resulting in a large percentage modulation, the demodulator may operate over the curved portion and serious distortion will result. A method of overcoming this is to use some type of tuning indicator so that tuning is done visually and not by ear.

A tuning indicator may be fitted in several ways:

(a) A meter may be fed from a rectifier connected to one of the i.f. amplifiers to a stage prior to limiting. This has the disadvantage that, since the response of the i.f. should be fairly flat over the required bandwidth, the meter does not readily indicate correct tuning. The receiver is, of course, tuned to maximum reading, and the meter reading gives some indication of the strength of the received signal.

(b) A good Q tuned circuit (tuned to 10·7 MHz) may be coupled, say, by mutual inductance, to an early i.f. stage (before limiting occurs). The voltage across the tuned circuit is then rectified and feeds a meter through, say, an emitter-follower. This is similar to (a), but the variations of meter reading can be made greater for a given degree of mistuning. The meter reading will also give an indication of the signal strength.

(c) A better way, and what appears simpler, is to use the output of the ratio detector. It was shown that the output from a balanced ratio detector was zero at the centre frequency, and varied both in the positive and negative directions with deviation. Thus, an indicator across the output of the demodulator can be used as a tuning indicator. This may be a centre-zero meter; the set is tuned for zero voltage. The disadvantage of this arrangement is that no indication of signal strength is given. The meter is preferably fed through a transistor so that it does not load the ratio detector.

(d) Indicator lamps may be used, one arrangement being given in figure 2.45. The two lamps L_1 and L_2 are fed from a symmetrical long-tailed pair consisting of Tr_2 and Tr_3, with current amplifiers Tr_1 and Tr_4 (forming

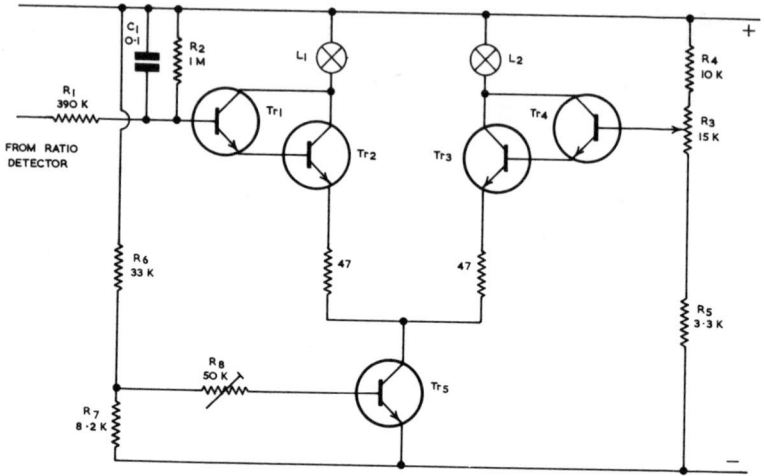

FIG. 2.45 F.M. TUNING INDICATOR

Darlington pairs). The output from the ratio detector is fed to the base of Tr_1, through resistor R_1, the capacitor C_1 removing any a.f. component. The base is also fed from the positive line through R_2 so that Tr_1 and Tr_2 are partly on. This is necessary because the voltage from the ratio detector is both positive and negative and requires the lamps to increase and decrease in brightness. Hence they must start with approximately half brightness, when there is zero voltage from the ratio detector. Due to Tr_2 and Tr_3 forming a long-tailed pair, if the current of Tr_1 is increased the current of Tr_2 is also increased, while that of Tr_4 and Tr_3 will be reduced. The lamps are adjusted so that they are equally bright, when there is zero voltage from the ratio detector, by means of the preset potentiometer R_3. The brightness of the two lamps can be varied by varying R_8, which controls the base current of Tr_5, which in turn controls the emitter currents (and hence lamp currents) of Tr_2 and Tr_3. This is a very sensitive indicator because it is easy to adjust the two lamps until they are equally bright.

In place of, or in addition to, a tuning meter, automatic frequency control can be used. Normally the a.f.c. is switched out and the set tuned approximately to the required station, and then the a.f.c. is switched on. The a.f.c. then automatically accurately tunes the receiver to the required station. To do this we require a control signal, which is approximately proportional to the error, and a device which makes the required changes in the tuning. The first signal is easily obtained (as we have seen) from the output of a balanced ratio detector. The correction in tuning is done by means of a variable capacitance diode (varactor or varicap diode) connected across the oscillator tuned circuit. A variable capacitance diode is one in which its capacitance changes as the reverse voltage is changed. The connection of such a diode to an oscillator is given in figure 2.46. Tr_1 is a combined oscillator-mixer, the oscillator tuned circuit being L_2 C_4. D_1 is the variable capacitance diode which,

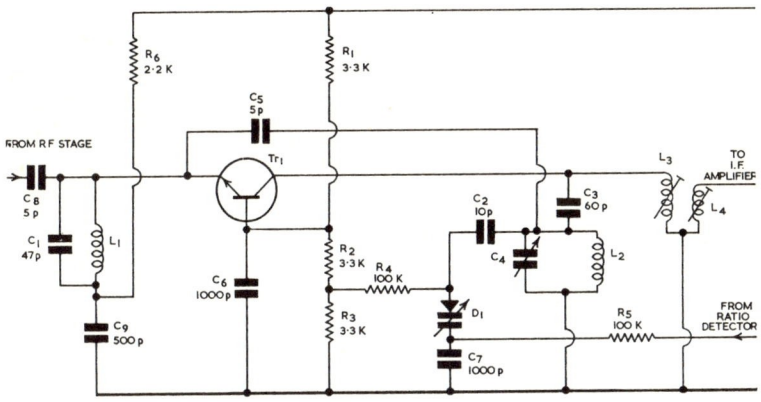

FIG. 2.46.   AUTOMATIC FREQUENCY CONTROL OF OSCILLATOR

from the point of view of a.c., is connected across C_4, since C_7 is of low reactance. As the voltage from the ratio detector varies in both directions, and the diode must at all times be reverse biased, a reverse bias voltage is necessary, which is obtained from across R_3 and forms part of the potential divider R_1, R_2 and R_3 across the supply. The high resistor R_4 is essential to prevent R_3 (of low value) being connected in parallel with the tuned circuit. The output from the ratio detector is fed through R_5 to the diode, R_5 and C_7 acting as a filter to remove the a.f. voltage. Thus, any error in tuning causes an output voltage to be fed to the diode in such a direction as to change the oscillator frequency to correct the tuning. Obviously, the circuit must be arranged so as to reduce the error and not increase it. Another circuit is given in figure 2.47, where L_1C_1 form the tuned circuit of the oscillator. Again D_1, the variable capacitance diode, is in parallel with C_1, the circuit being completed through C_2, C_4 and C_3 as regards a.c. The output from the ratio detector is filtered by R_3 C_5 (to remove a.f.) and then fed to D_1. Resistor R_2 isolates C_5 from the tuned cicruit. The fixed bias is now obtained from the zener diode D_2, fed from the positive line through R_1. It is only necessary to alter the tuning of the oscillator as the frequency correction required is only small, say ± 250 kHz. Although ideally the station should first be tuned in without a.f.c., in some receivers the a.f.c. cannot be switched off. In this case the set is tuned to approximately the correct setting and then the a.f.c. pulls the set into tune. A.F.C., of course, also compensates for any drift in the oscillator frequency.

PRESET TUNING USING VARIABLE CAPACITANCE DIODES

Most f.m. receivers are made with continuously variable tuning, usually by means of a ganged variable capacitor, but variable inductors (permeability tuning) are sometimes used. It is convenient, however, to be able to select a number of stations by means of push-buttons. At the high frequencies concerned it is difficult to select preset tuned circuits by means of switches operated

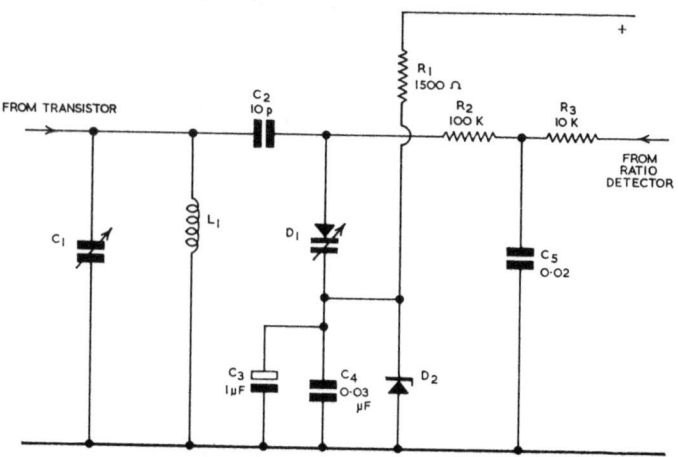

FIG. 2.47. CONTROL OF OSCILLATOR BY A.F.C. VOLTAGE

by push-buttons so the ideas of using variable capacitance diodes was introduced. Preset potentiometers are selected by the push-buttons and the voltage
then used to vary the capacitances of the diodes, which then control the tuning
of the various circuits. As the push-buttons are only carrying d.c. they can be
placed in the most convenient place in the cabinet, and they can easily be
extended outside the cabinet for remote control if required. A basic circuit
of such a receiver is given in figure 2.48. Since it is now required to tune the
receiver over its full range all the tuned circuits must be varied by variable
capacitance diodes.

The circuit shown has three tuned circuits; the input and output of the
r.f. amplifier, and the oscillator. Considering the input tuned circuit L_2C_1,
the variable capacitance diode D_1 is placed in parallel with C_1 as regards
a.c., the capacitor C_2 being a d.c. blocking capacitor. The reverse voltage
(a positive voltage) is fed through R_1. C_1 is a trimming capacitor used to align
the circuits in the first instance. The output tuned circuit is L_3C_4 and the
variable capacitance diode D_2 is in parallel with C_4. Again, C_4 is a trimming
capacitor only. D_2 is fed with control voltage through R_2, which prevents the
tuned circuit being shorted by the control voltage source. The oscillator
tuned circuit is L_5 with trimming capacitor C_8, now tuned by variable capacitance diode D_3 which is fed through the isolating choke L_6. No bias is required
on these diodes as the control voltage is in one direction only (the reverse
bias direction). A stable voltage is obtained from the supply by the use of a
zener diode D_5 and smoothing circuit R_6, C_{10}. The four preset potentiometers
are fed from this supply and are selected by push-button switches S_1 to S_4. C_{11}
acts as additional smoothing and r.f. by-pass. If continuous tuning is also
required then R_8 (for example) can be made a variable potentiometer and a
suitable scale provided for it to read frequency. Alternatively, a variable
potentiometer can be used with a meter connected across the output of this
control (*i.e.* to measure the control voltage), the scale being suitably calibrated in MHz. The meter may also be used to indicate the frequency of the
preselected stations. Any number of buttons can, of course, be used. Because
the accuracy of tuning by variable capacitance diodes is not good, a.f.c. is
also used, D_4 being the control diode. This is fed with fixed reverse bias from
the potential divider R_4 R_5 and fed with a.f.c. control voltage through the
22 kΩ isolating resistor. D_4 is connected across L_5 as regards a.c. and therefore
controls the oscillator frequency.

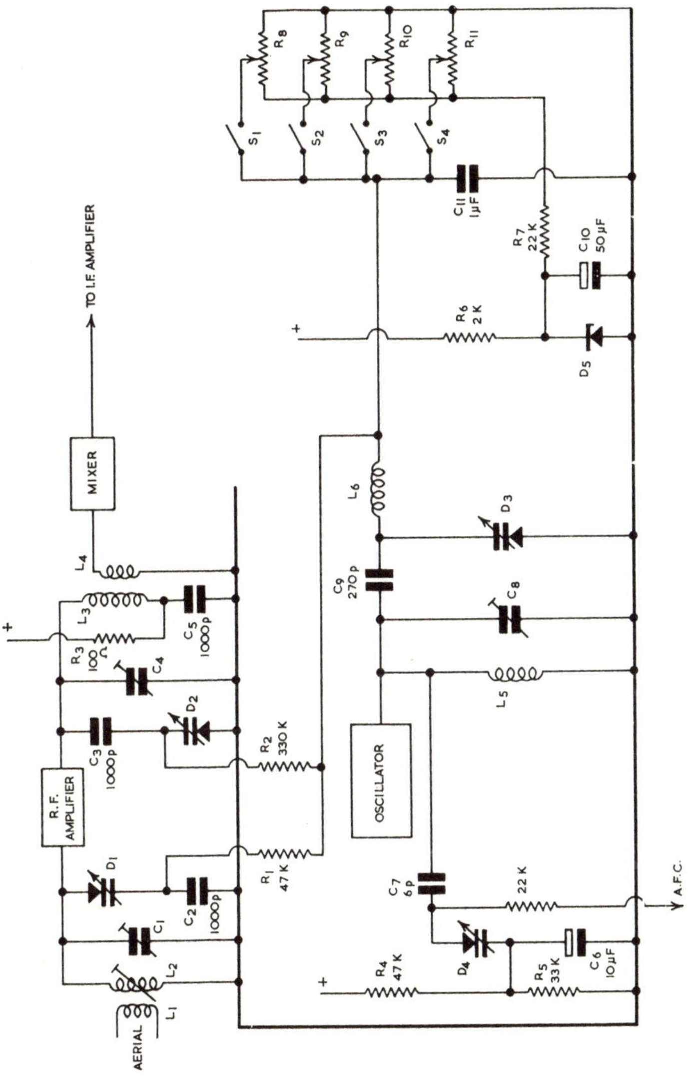

FIG. 2.48. PRESET TUNING USING VARIABLE CAPACITANCE DIODES

AM/FM RECEIVERS

Although this chapter is concerned with f.m. receivers, it is useful to say something about the arrangements used in combined a.m./f.m. receivers as these are common. In many ways the simplest arrangement is to use separate circuits up to, and including, the demodulators, but this is expensive and usually some attempt is made to use some of the transistors for both a.m. and f.m. Generally, the r.f. stage, oscillator and mixer on f.m. are separate from the corresponding a.m. sections, but the i.f. stages are commoned. The two i.f. transformers, one tuned to 10·7 MHz for f.m. and the other to 470 kHz for a.m., are simply connected in series as in figure 2.49. C_1, L_1, C_2, L_2 form the f.m. i.f. transformer, and C_3, L_3, C_4, L_4 the a.m. transformer. At 10·7 MHz the reactance of C_3 and C_4 will be so low that effectively the a.m. transformer is shorted out. Similarly, at 470 kHz the reactances of L_1 and L_2 are low, and have little effect on the circuit operation. R_1 and R_2 provide bias for Tr_2, and C_5 completes the path for the i.f. Separator demodulators are obviously necessary and must be switched to the a.f. amplifier as required. The number of i.f. stages used may be different for a.m. and f.m., a greater number being used on f.m. This can be achieved by using the a.m. oscillator as an i.f. stage on f.m.

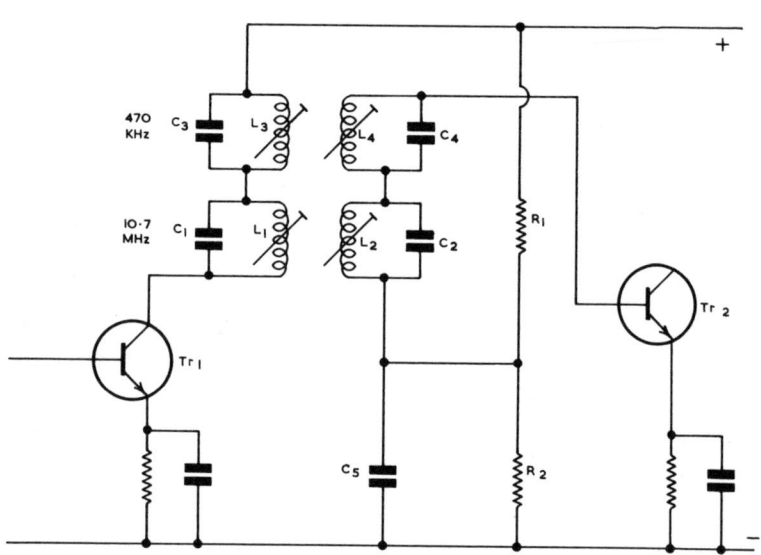

FIG. 2.49. A.M.-F.M. I.F. AMPLIFIER

CHAPTER 3

A.F. POWER AMPLIFIERS

Iɴ *Radio Servicing*, Volume 2, audio frequency amplifiers were described, including those using complementary transistors, which were suitable for small radio receivers with an output of, say, up to 1 watt. It is now intended to deal with amplifiers of much larger output as used in hi-fi equipment.

Because of the low efficiency of class-A operation, amplifiers operating class-B are nearly always used, although there is at least one class-A commercial amplifier. Transformers are not generally used; some type of transformerless circuit is used, most being based on the complementary transistor amplifier described in Volume 2. A knowledge of such circuits will be assumed.

For a given loudspeaker impedance the only way of increasing the power output is to increase the voltage across the speaker, which means a greater supply voltage to the transistors. The greater voltage also means a greater current through the loudspeaker and output transistors. This, in turn, means a larger drive current to the bases, and hence a larger driver stage. The driver stage must provide both half-cycles so it must operate class-A. Its standing current must be greater than the peak base current required. Therefore, as the output power of the amplifier increases, the power rating of the driver stage becomes large. Since it takes a large current its load resistance must be low, and so the gain is low. At some stage this arrangement becomes unsatisfactory, and the difficulty is overcome by using current amplifier stages between the driver stage and the output transistors. A basic circuit is given in figure 3.1. Tr_1 and Tr_2 are the complementary output transistors feeding the loudspeaker through the blocking capacitor C_1. Tr_3 and Tr_4 are the current amplifiers connected as emitter-followers. Apart from the small current in R_3 and R_4 the emitter current of Tr_3 is the base current of Tr_1 and the emitter current of Tr_4 is the base current of Tr_2. Thus the base current required by Tr_3 and Tr_4 is the base current of Tr_1 or Tr_2 divided by the current gain of Tr_3 or Tr_4, *i.e.* the current gain β or h_{fe}. Since the current gain can easily be 50-100 it makes the problem of the driver stage Tr_5 relatively easy. R_2 is the load resistor for the driver stage which in previous circuits has been shown as being returned to the junction of the emitter of the two output transistors. This is what is sometimes called "bootstrapping" and reduces the current variations required in the driver stages. In this case, as the current has been so greatly reduced, bootstrapping is not necessary and it can be fed from the positive line. R_1 provides bias for all four transistors Tr_1 to Tr_4. As explained in *Radio Servicing*, Volume 2, the use of a resistor for this bias is not very satisfactory due to temperature variations and supply voltage changes. It may be replaced by a thermistor or thermistor/resistor combination, which compensates for temperature changes but not for supply voltages changes. Alternatively, a diode or a number of diodes in series may be used, these compensating to a large extent for temperature and supply voltage changes. Diodes have the disadvantage that their voltage drop cannot be varied, and hence a transistor may be used and connected as in figure 3.2. As the voltage across this circuit rises the base voltage rises and causes the transistor to conduct and oppose the voltage rise. The characteristic is, in fact, similar to that of a diode, but the voltage across it can be varied by varying R_1. If the temperature rises then the base-emitter voltage drops, for a given base current, and so compensates for the reduced base-emitter drop of the output transistors. It is a good idea to mount the transistor on the same heat sink as the output transistors, so that they are all at approximately the same temperature. Such a transistor circuit results in an almost constant voltage source. In a particular case, changing the current from 8 to

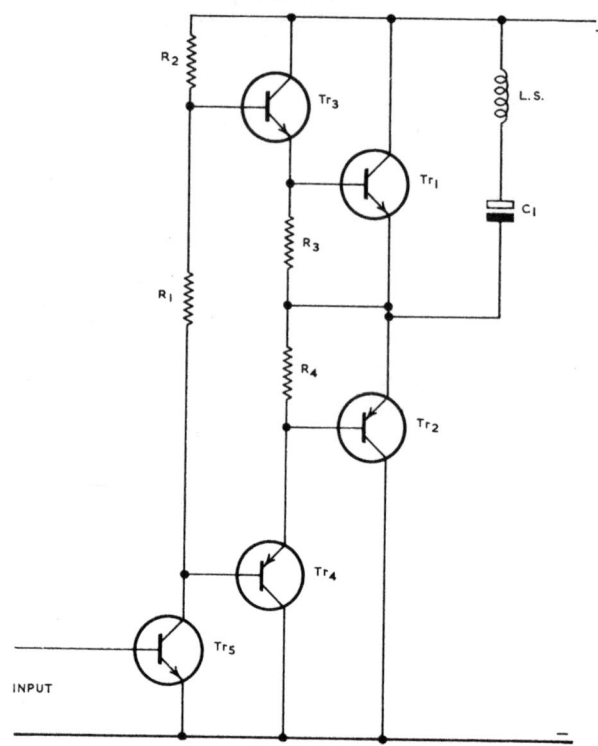

FIG. 3.1. OUTPUT STAGE WITH CURRENT AMPLIFIERS

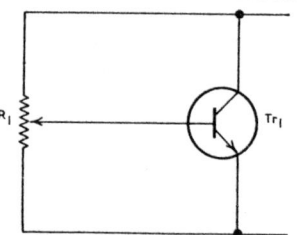

FIG. 3.2. CONSTANT VOLTAGE TRANSISTOR CIRCUIT

16 mA (100% increase) only changed the voltage from 2·65 to 3 volts (13% increase).

One of the problems of transistor amplifiers is to maintain the correct operating conditions so that the junction of the emitters of the output transistors is at approximately half the supply voltage. If the voltage at this point drifts, due to temperature or ageing, then the maximum power output will be reduced. A commonly-used basic circuit is given in figure 3.3. Tr_1 and Tr_2 are the main output transistors and are fed by the current amplifiers Tr_3 and Tr_4. It might at first appear that there is no reason for R_3 and R_4 as they would only shunt off part of the current from Tr_3 and Tr_4. Their function is to carry any leakage current of Tr_1 and Tr_2. Considering Tr_1, even with no emitter current there will be a small leakage current through the collector-base junction, and this current will tend to flow out of the base. Without the presence of R_3

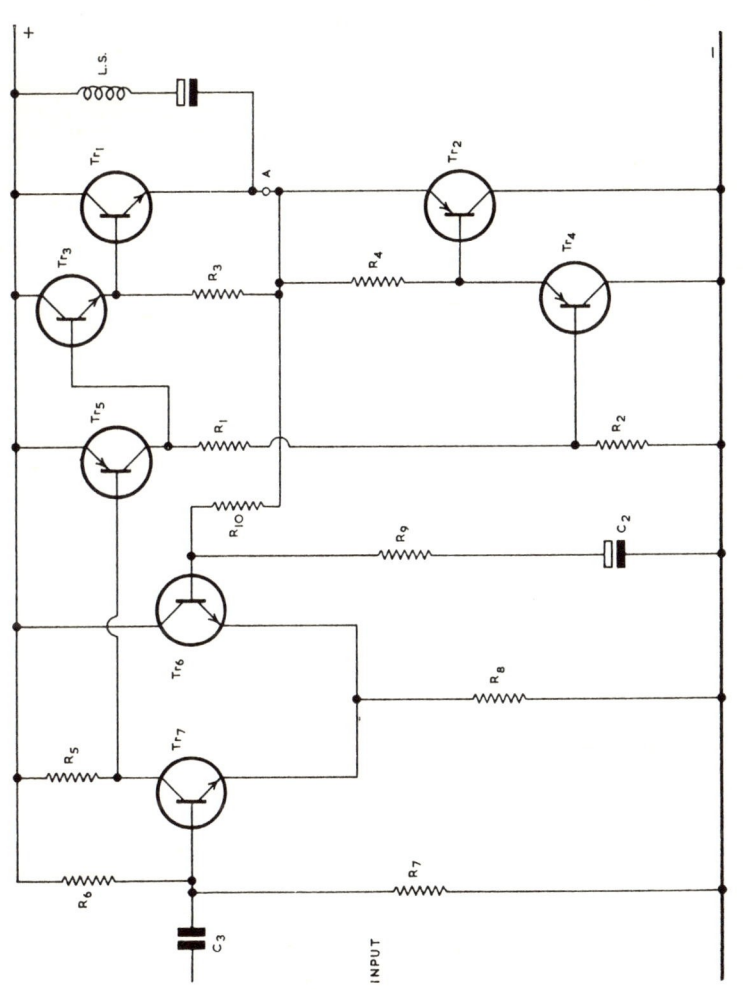

FIG. 3.3. COMPLEMENTARY OUTPUT STAGE WITH D.C. AND A.C. FEEDBACK

there is no external path for this current to flow, and it must flow through the base-emitter junction. Due to transistor action this will cause a greater collector current to flow, hence a greater base-emitter current, and so on. This results in a relatively large leakage current flowing, particularly at high temperatures. By adding R_3, (of fairly low value) most of the collector-base leakage current will flow in R_3 and not in the base-emitter junction.

Tr_3 and Tr_4 are fed from the driver stage Tr_5 which has a load resistor R_2. Resistor R_1 provides bias of Tr_1 to Tr_4, but would normally be replaced by a thermistor, diode or transistor as already explained. On its own, this circuit would be unstable because any change in the standing current of Tr_5 would alter the voltage across R_2. Therefore, the base voltages of Tr_3 and Tr_4 would be altered and the voltage at point A, which should be at half the supply voltage. To overcome this some d.c. negative feedback must be used. In this circuit it is arranged by using Tr_6 and Tr_7 as a long-tailed pair, with the common emitter resistor R_8 (which may be replaced by a constant-current circuit). Bias for Tr_7 is provided by R_6, R_7 and the voltage on the base of Tr_7 is arranged to be approximately half the supply voltage. The a.f. input is fed through C_3 to the base of Tr_7, where it is amplified, and the voltage across the collector load R_5 is fed to the driver stage Tr_5. The base of Tr_6 is returned to point A through resistor R_{10}. Now, it will be remembered that in a long-tailed pair, unless one of the transistors is cut off, the two base voltages must be at approximately the same voltage. They cannot differ in voltage by more than twice the base-emitter drop, say a voltage difference of 1 volt, or one transistor will be cut off. Thus the circuit will sort itself out so that the potential at A, and hence the base voltage of Tr_6, will be approximately the same as that at the base of Tr_7, which is at half the supply voltage. For example, suppose that the voltage at A rises and therefore the base voltage of Tr_6 rises, so that Tr_6 current rises. This will cause the voltage across R_8 to rise and so reduce the base-emitter voltage of Tr_7, and hence its current. This, in turn, will cause a reduction in voltage across R_5, and a reduced base-emitter voltage on Tr_5. Thus the current of Tr_5 will be reduced, so reducing the voltage across R_2. This reduces the voltages on the bases of Tr_3 and Tr_4, and, since they are operating as emitter-followers, the emitter voltages. This reduces the base voltages of Tr_1 and Tr_2, and tends to lower the potential of point A until equilibrium is reached. A.C. feedback is also used from point A, to reduce distortion. As regards a.c., R_9 and R_{10} form a potential divider from point A (C_2 having a negligible reactance). Thus the amount of a.c. feedback can be varied by varying the ratio of R_9 to R_{10}, the value of R_{10} having little effect on the d.c. feedback.

An alternative current-amplifier circuit is shown in figure 3.4. In this case the current amplifier transistors are complementary to the main output transistors. Tr_5 is the driver transistor, with load resistor R_2 and bias resistor R_1. Suppose the voltage across R_2 is reduced owing to a reduction in Tr_5 collector current. This will cause a rise in base voltage on Tr_3 and, because it is an emitter-follower, the emitter voltage will rise and provide some of the output power. However, as the collector current increases it flows out of the base of Tr_1 (apart from the small current in R_3) and will therefore cause a much greater collector current to flow in Tr_1. Because of the current gain of Tr_1 it is this transistor which supplies most of the output current. One may assume that if the emitter of Tr_3 does not follow the base voltage, it causes Tr_1 to conduct more until the emitter is almost at the same potential as the base. Tr_4 and Tr_2 operate in a similar manner. The arrangement of figure 3.3 is the more common. We have now covered the main points as regards fully complementary output stages. Obviously, many detailed variations are possible,

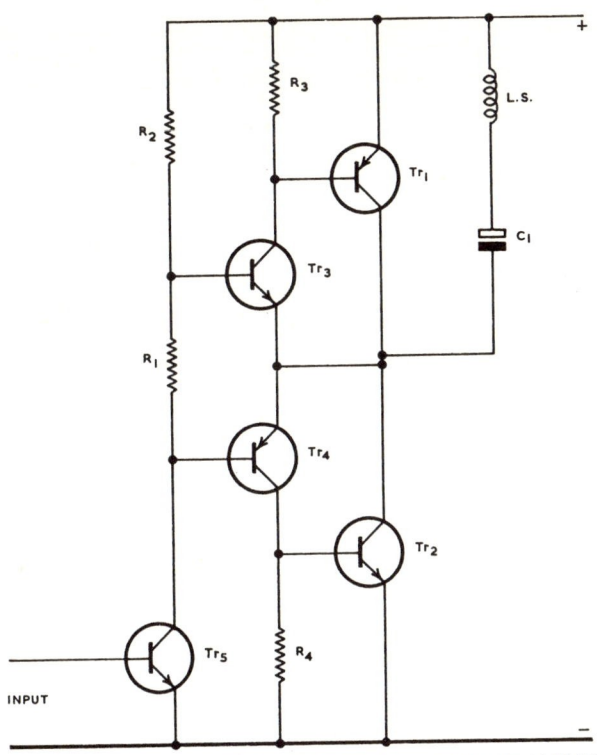

FIG. 3.4. USE OF COMPLEMENTARY TRANSISTORS AS CURRENT AMPLIFIERS

e.g. the load resistor of the driver stage may comprise a constant-current circuit rather than a resistor.

It will be noted that the output transistors in the circuits described have all been connected as common-collector circuits, *i.e.* emitter-followers. This configuration has the advantage that if the base is driven with an undistorted voltage then the output voltage must also be undistorted (to the first approximation). The current gain of power transistors varies greatly with collector current and may change, for example, from 100 at small currents to 20 at large currents. The only effect of this is to cause the base current to be distorted. Therefore the base should be driven from a low impedance source so that as far as possible the voltage is undistorted. The disadvantage of this configuration is that it has no voltage gain; the driver stage must therefore produce a voltage equal to the output voltage. The alternative configuration is common-emitter (common base is impractical as there is no current gain). This arrangement has voltage gain, and therefore the voltage output required by the driver stage is considerably reduced. However, if the base is driven with an undistorted current, distortion will occur as a result of the change in current gain over the cycle. If driven with an undistorted voltage then the output will be distorted due to the curvature of the collector current-base voltage characteristic. Some of the distortion can be removed by negative feedback, but common-emitter stages do not appear to be normally used in large power amplifiers.

Although a complementary output stage may sound ideal, there are two drawbacks. First, it requires a matched pair of complementary transistors; such devices are manufactured as being matched but if the characteristics are studied it will be found that generally they are far from being matched. However, the use of emitter-follower stages and overall negative feedback largely reduces the effects of mismatching. Secondly, *n-p-n* silicon power transistors are cheaper than their *p-n-p* counterparts of similar rating. As a result of this, what are called quasi-complementary amplifiers are commonly used. An example of such a circuit is given in figure 3.5. The two output transistors Tr_1 and Tr_2 are now the same—both *n-p-n*. They could, of course, be *p-n-p*, but

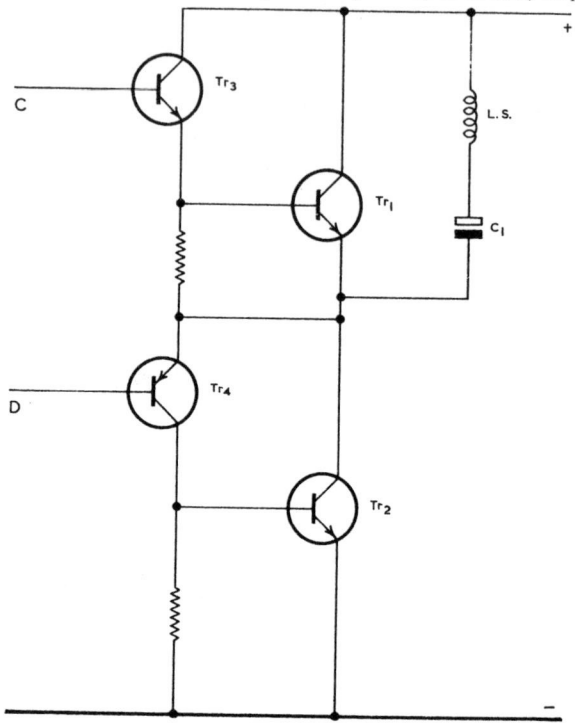

FIG. 3.5.   QUASI-COMPLEMENTARY OUTPUT STAGE

there is little point as they are more expensive. One of the current amplifier Tr_3 is connected as an emitter-follower, and therefore no phase reversal takes place. However, Tr_4 is connected as a common-emitter stage, and there is a 180° phase shift between the input to the base and the feed to the base of Tr_2. Thus, if the same signal is fed to C and D, antiphase voltages will be fed to the bases of Tr_1 and Tr_2 as is required.

A complete quasi-complementary stage is shown in figure 3.6 where Tr_1, Tr_2, Tr_3 and Tr_4 are connected as in figure 3.5. The driver transistor is Tr_5 with its load resistor R_2. R_1 provides the bias for transistors Tr_1 to Tr_4 and would normally be replaced by diodes, a transistor or a thermistor. Instead of taking the upper end of the load resistor R_2 to the positive line a bootstrap circuit is used. As regards a.c., the upper end of R_2 is connected to point A through the low reactance of capacitor C_3. The voltage across R_2 is, therefore, applied directly between the base and emitter of Tr_3 and Tr_4. The d.c. is fed to Tr_5 through R_5. Tr_6 feeds the driver stage from its load resistor R_9. Bias for Tr_6 is obtained from R_6 and R_7 and the bias voltage is arranged to be approximately half the supply voltage. The emitter of Tr_6 is taken to point A through R_{10}.

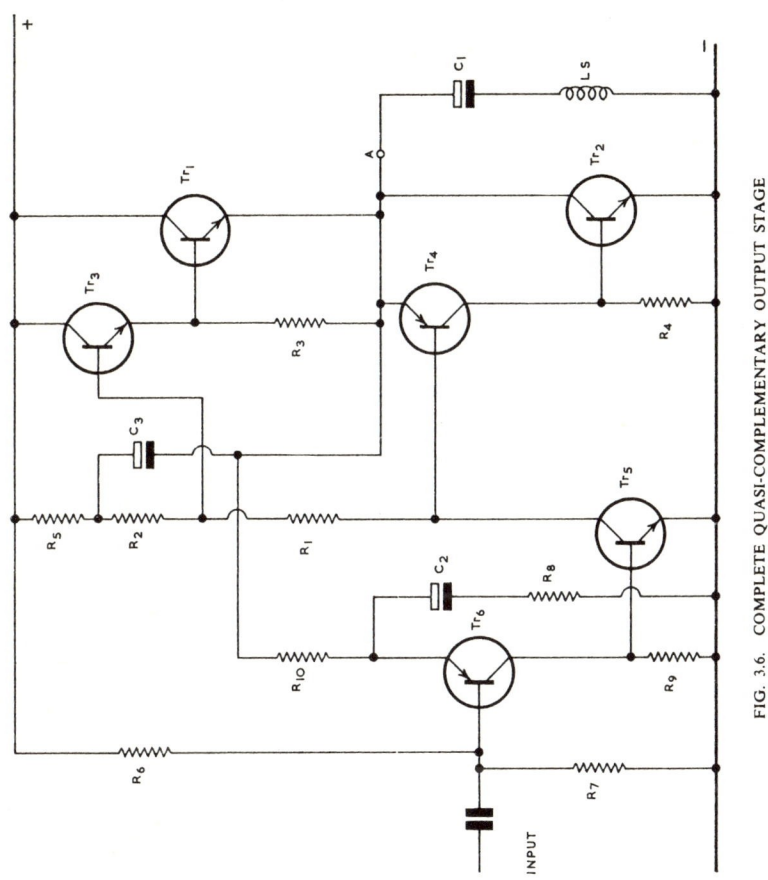

FIG. 3.6. COMPLETE QUASI-COMPLEMENTARY OUTPUT STAGE

Since the base-emitter voltage of Tr_6 can only be a fraction of a volt, the circuit must sort itself out so that the potential at A is approximately that on the base of Tr_6, which, as stated earlier, is made about half the supply voltage. Thus A is at half the supply voltage as required. If the voltage at A rises it causes an increase in base-emitter voltage on Tr_6, and an increase in collector current. This causes an increased drop across R_9, which, therefore, increases the current in Tr_5, this causing increased drop across R_2 so that the bases of Tr_3 and Tr_4 are reduced in voltage. This causes the emitter of Tr_3 to be reduced as it operates as an emitter-follower. This, in turn, causes the base, and hence emitter voltage, of Tr_1 to be reduced, so tending to compensate for the original rise in voltage, i.e. there is a negative feedback loop. If Tr_4 is considered the drop in base voltage will cause an increase base-emitter voltage, and increased collector current. This causes an increased voltage across R_4 and increases the current of Tr_2, which will again cause point A to be reduced in voltage.

There will be a.c. feedback through R_{10}, but the amount is controlled by the value of R_{10} relative to R_8, since these two resistors form a potential divider, C_2 having a negligible reactance.

An unusual arrangement is shown in figure 3.7 which is not quasi-complementary but uses Tr_5 to produce the phase splitting. Tr_3 and Tr_1 form a Darlington n-p-n pair and similarly Tr_4 and Tr_2 form another Darlington pair. For correct operation the bases of Tr_3 and Tr_4 must be driven in antiphase. Transistor Tr_5 has equal collector and emitter loads, R_5 and R_6. If the current is increased the collector voltage will fall, and that of the emitter will rise. Thus collector and emitter voltages are 180° out of phase, and so are suitable for feeding the bases of Tr_3 and Tr_4. Tr_3 is fed directly, since the steady collector voltage happens to be of suitable value. Tr_4 must be fed through a capacitor C_2 to remove the steady voltage on the emitter, which is not of the correct value. Bias is fed to Tr_4 from R_1 and R_2. Tr_6 is a preamplifier driving the phase-splitter stage Tr_5. Tr_5 will have no voltage gain due to the 100% feedback across R_6. D.C. and a.c. negative feedback are obtained by connecting R_9 from point A to the emitter of Tr_6. Should the voltage at point A rise it will cause the emitter of Tr_6 to rise, and, assuming its base voltage is constant, will reduce the current in Tr_6. This causes the collector voltage to rise and increase the base voltage and collector current of Tr_5. Thus, the voltage across R_5 is increased, causing the base of Tr_3 to drop in voltage. As a result the base and emitter of Tr_1 fall, so compensating for the initial rise of voltage at point A. There is both a.c. and d.c. feedback which are controlled by the ratio of R_9 to R_8.

In setting up these amplifiers the standing current and the voltage at point A (the junction of the two transistors feeding the loudspeaker) should be correct. The bias must be set so that a suitable standing current flows with no signal applied. This is necessary to prevent cross-over distortion. The value of the standing current will depend on the size of amplifier. If it is too small then some cross-over distortion is likely, and if too large it will cause excessive heating of the transistors. It is best set to the manufacturer's figure if available. The method of varying the bias will depend on the arrangements used. If diodes are used little adjustment is possible. The simplest is the transistor method (figure 3.2) since the bias is varied by the preset potentiometer. There will often be some means of adjusting the potential of point A, the method depending on the circuit. For example, in figure 3.3 R_6 or R_7 may be made preset, and the same applies to figure 3.6. The voltage at point A should be half-way between the two line voltages so that the output load (point A) can vary equal amounts in the two directions. If this is not the case then clipping will occur in one direction before the other, and the maximum output power will be reduced.

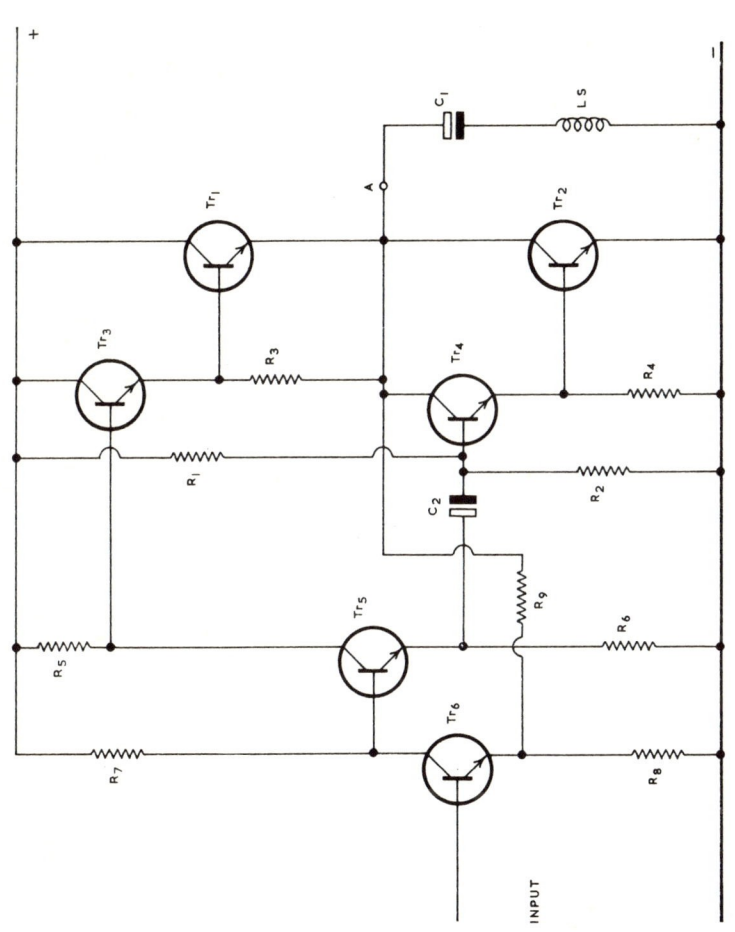

FIG. 3.7. OUTPUT STAGE WITH PHASE SPLITTER

EFFECT OF LOUDSPEAKER IMPEDANCE

The effect of loudspeaker impedance is rather important as it determines the maximum output and also the transistor dissipation. Figure 3.8 shows the conditions on an output transistor, assuming perfect class-B operation. When

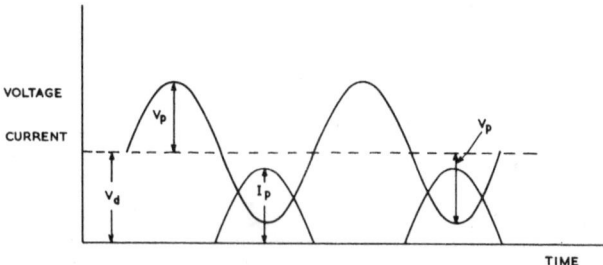

FIG. 3.8. VOLTAGE AND CURRENT OF OUTPUT TRANSISTOR

the transistor is conducting there is a half-cycle of collector current of peak value I_p. The d.c. supply to one transistor is V_d and the collector voltage varies about this value. When collector current flows the collector voltage varies in a sinusoidal manner of peak value V_p, and as shown, the collector voltage is reduced; under full output conditions it will drop to almost zero at the instant of maximum current, i.e. $V_p = V_d$. During the other half-cycle there is no collector current, but the other half-cycle of output (produced by the other transistor) causes the collector voltage to rise above V_d to a maximum value of $V_d + V_p$ and, under maximum output conditions, V_p is nearly equal to V_d, hence the peak collector voltage becomes $2V_d$. It is important to note this because the transistor must stand this voltage without breakdown. Note also that V_d is the supply to one transistor and when two transistors are in series (as in the circuits described in this book) then $V_d = \frac{1}{2}V_s$ where V_s is the total supply voltage. In other words, each transistor must withstand a voltage V_s.

It is shown in Appendix 1 (equation A.2, page 197) that the power dissipated in both transistors is

$$W = \frac{2V_dV_p}{\pi R} - \frac{V_p^2}{2R} \qquad (3.1)$$

where R is the impedance of the load which is assumed resistive.

Now $\dfrac{2V_dV_p}{\pi R}$ is the power input to the transistors

and $\dfrac{V_p^2}{2R}$ is the power output to the load

the difference, of course, must be the transistor dissipation.

$$
\begin{aligned}
\text{The efficiency} \quad &= \quad \frac{\text{power output}}{\text{power input}} \\[6pt]
&= \quad \left(\frac{V_p^2}{2R}\right)\Big/\left(\frac{2V_dV_p}{\pi R}\right) \\[6pt]
&= \quad \frac{\pi V_p}{4V_d} \qquad\qquad (3.2)
\end{aligned}
$$

Under maximum output conditions we will assume that $V_p = V_d$ (in practice V_p must be slightly less than V_d) hence the efficiency under maximum output

conditions is $\dfrac{\pi}{4}$ or 78·5%.

One might expect that the dissipation in the transistors would be a maximum under conditions of full output but, as shown in Appendix 1, this is not the case, and maximum dissipation occurs when $V_p = 0.636V_d$.

Substituting this in equation (**3.2**), the efficiency is then

$$\frac{\pi \times 0.636V_d}{4V_d} = 50\%$$

Figure 3.9 shows how the input, output, transistor dissipation and efficiency vary as the output voltage V_p varies.

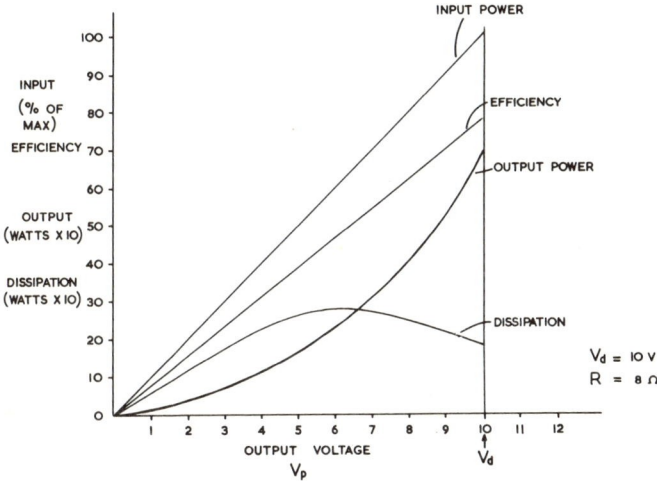

FIG. 3.9. RELATIONSHIP BETWEEN OUTPUT VOLTAGE AND INPUT POWER, EFFICIENCY, OUTPUT POWER AND TRANSISTOR DISSIPATION

Consider an amplifier feeding an 8 ohm load and with a supply voltage (per transistor) of $V_d = 10$ volts.

(a) UNDER MAXIMUM POWER OUTPUT CONDITIONS $V_p = V_d = 10$ V

Power input $= \dfrac{2V_dV_p}{\pi R} = \dfrac{2 \times 10 \times 10}{\pi \times 8} = 7.958$ watts

Power output $= \dfrac{V_d{}^2}{2R} = \dfrac{10^2}{2 \times 8} = 6.250$ watts

Power dissipation $= 1.708$ (the difference between input and output).

(b) UNDER CONDITIONS OF MAXIMUM DISSIPATION $V_p = 0.636 \ V_d = 6.36$ V

Power input $= \dfrac{2 \ \times \ 10 \ \times \ 6.36}{\pi \ \times \ 8} = 5.061$

Power output $= \dfrac{6.36^2}{2 \ \times \ 8} = 2.528$

Power dissipated $= \underline{2.533}$

Thus it is seen that under these conditions the transistor dissipation is considerably greater than at maximum output, actually 48% greater.

(c) EFFECT OF INCREASING V_d. Suppose V_d is doubled to 20 V

Under maximum output conditions when $V_p = V_d = 20$ V

Power input $= \dfrac{2 \ \times \ 20 \ \times \ 20}{\pi \ \times \ 8} = 31.83$ watts

Power output $= \dfrac{20^2}{2 \ \times \ 8} = 25.00$ watts

Power dissipation $= \underline{6.83 \text{ watts}}$

If these are compared with (a) it is seen that all the powers are quadrupled. For a given load impedance the only way of increasing the power output is by increasing V_d.

(d) EFFECT OF INCREASED LOAD RESISTOR. Assume $V_d = 10$ V and $R = 16 \ \Omega$

Under maximum power output conditions when $V_p = V_d = 10$ V

Power input $= \dfrac{2 \ \times \ 10 \ \times \ 10}{\pi \ \times \ 16} = 3.979$ watts

Power output $= \dfrac{10^2}{2 \ \times \ 16} = 3.125$ watts

Power dissipation $= \underline{0.854 \text{ watts}}$

If these are compared with (a) it is seen that all the powers are halved and, most important perhaps, the output power is halved.

(e) EFFECT OF DECREASED LOAD RESISTOR. Assume $V_d = 10$ V and $R = 4 \ \Omega$

Under maximum power output conditions when $V_p = V_d = 10$ V

Power input $= \dfrac{2 \ \times \ 10 \ \times \ 10}{\pi \ \times \ 4} = 15.915$ watts

Power output $= \dfrac{10^2}{2 \ \times \ 4} = 12.5$ watts

Power dissipation $= \underline{3.415 \text{ watts}}$

If this is compared with (a) it is seen that all the powers are doubled and, what is most important, that the transistor dissipation is doubled. Hence, if an amplifier designed for an 8 ohm load is operated with a 4 ohm load it is likely to be damaged due to excessive transistor dissipation. Damage might also be done to the power supply since its output is also doubled. The transistor and power supply currents are also doubled. The current may, of course, be limited by a protection circuit in the amplifier.

One might expect that at least the same output could be obtained without overloading the transistors. This is not the case. If we consider a power output of 6·25 watts as at (a) the voltage for this output with a 4 Ω load is 7·07 V, hence:

$$\text{Power input} = \frac{2 \times 10 \times 7\cdot07}{\pi \times 4} = 11\cdot252 \text{ watts}$$

$$\text{Power output} = \frac{7\cdot07^2}{2 \times 4} = 6\cdot248 \text{ watts}$$

$$\text{Power dissipation} = \underline{5\cdot004} \text{ watts}$$

The dissipation is much greater than the corresponding figure at (a) of 1·708 watts and even much greater than the worst condition with an 8 ohm load as at (b) of 2·533 watts.

The worst condition with a 4 ohm load results in a dissipation of 5·066 watts (twice that for an 8 ohm load).

(f) SHORT CIRCUIT CONDITION. In this case $R = 0$, and $V_p = 0$ $V_d = 10$

The previous expressions cannot be used because they give infinity for the powers. In practice, the current output will be limited, probably by the driver stage. We must now use the other expression in the Appendix 1 (equation A.1). It will be assumed that the current is limited to the maximum value which results in case (a) with an 8 ohm load. The current is 1·25 A.

$$\text{Power input} = \frac{2V_d I_p}{\pi} = \frac{2 \times 10 \times 1\cdot25}{\pi} = 7\cdot958$$

$$\text{Power output} = 0\cdot000$$

$$\text{Power dissipation} = \underline{7\cdot958}$$

It is therefore seen that the dissipation is very high and the amplifier can easily be damaged. In this example the current was limited to normal full output current but, in practice, it would be higher, particularly without a protection circuit, and hence serious damage can be done. The power dissipation is high because the transistor is passing current with full voltage across it, whereas, under normal conditions (figure 3.8), it is seen that when the current is flowing the transistor voltage is lower than V_d.

Overload protection will now be considered. It has been shown that if a speaker of too low an impedance is used the transistors can easily be overloaded, and that if the overload is maintained they will overheat and be damaged. The worst condition is when the loudspeaker leads are short circuited. Some protection is necessary as this is a fault that can easily occur when loudspeakers are separate from the amplifier (which is common in hi-fi equipment). The simplest protection is a fuse which will usually operate against short circuits in the equipment and short-circuited loudspeaker leads. In these cases the current rises to a high value and blows the fuse. A small circuit-breaker may be used instead, but is more expensive. It is more difficult to protect against an overload which does not result in a very excessive increase in collector current. One way is to detect the temperature of the transistors, usually by a thermistor which changes its resistance rapidly with temperature. If excessive temperature is detected there are two possible actions; remove the h.t. supply or remove the drive to the output transistors. Since the output stage is operated class-B, if the drive is removed the collector current will fall to a small value—its standing current. When the temperature falls then the supply voltage or drive can automatically be restored. The drive can be stopped by removing the supply voltage to the driver stage. If an electronic stabilizer is used to feed the amplifier then it is comparatively easy to arrange for the excess temperature to remove the supply. Excessive current trips can, of course, be fitted to the voltage stabilizer instead of using a fuse.

An alternative is to use some form of electronic limiter; a basic circuit is given in figure 3.10. Tr_1 is the output transistor, and a low value resistor R_1 is placed in its emitter circuit. Tr_2 is the current amplifier driving Tr_1, and

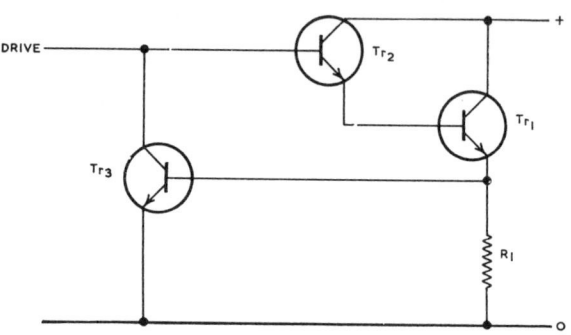

FIG. 3.10. OVERLOAD PROTECTION CIRCUIT

Tr_3 is the limiter. Under normal operating conditions the voltage across R_1 is not sufficient to turn Tr_3 ON and so it has no effect. If excessive current flows, the voltage across R_1 causes Tr_3 to conduct and so tends to short out the drive voltage. Thus the circuit prevents the current rising above some preset value. Only a half of the amplifier is shown, the other being identical. If we consider the I_c—V_{ce} characteristics of an output transistor it has been found that there is a prohibited area where the transistor must not operate or what is called **secondary breakdown** occurs. This area is shown in figure 3.11. The dotted line indicates the effect of the limiter Tr_3. It will be seen that the effect is not good because it allows the transistor to go into the prohibited area, for large values of V_{ce}, and limits the current to an unnecessarily low value at small

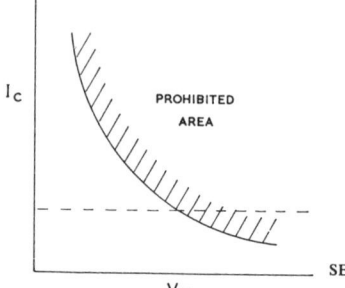

FIG. 3.11
SECONDARY BREAKDOWN AREA OF TRANSISTOR

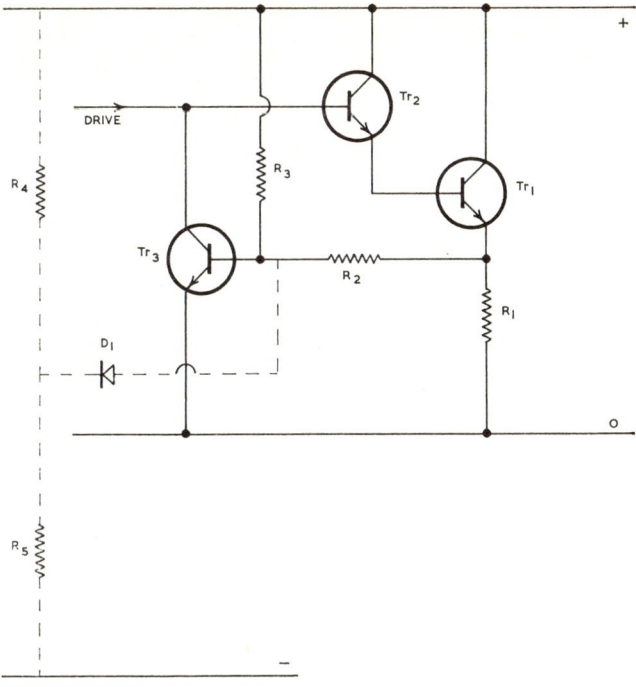

FIG. 3.12. IMPROVED OVERLOAD PROTECTION CIRCUIT

values of V_{ce}. This operation can be improved by connecting the circuit as in figure 3.12. Again, only one side of the amplifier is shown. R_1 operates as before but now the base is also fed from a potential divider R_2 R_3 which is connected across the output transistor. Thus, the voltage on the base of Tr_3 is the sum of a fraction of the transistor voltage and that across R_1, the latter being proportional to the collector current of Tr_1. The resulting characteristic is now shown by the full line of figure 3.13. Thus, when V_{ce} is low, the current that can flow is increased. The limiter now prevents the transistor operating in the prohibited area and makes fair use of the remaining area. Improved results can be obtained by connecting a diode D_1 to an artificial centre line produced by two equal resistors R_4 and R_5 across the supply lines. When V_{ce}

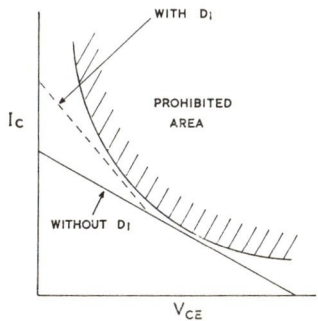

FIG. 3.13. CHARACTERISTIC OF IMPROVED OVERLOAD PROTECTION CIRCUIT

is low the junction of R_3 R_2 will go in a positive direction and, at some point, D_1 will conduct, so tending to reduce the voltage on the base of Tr_3. Thus to make Tr_3 conduct under these conditions, a larger current must flow in R_1. The new characteristic is shown dotted in figure 3.13 when it will be seen it makes better use of the non-prohibited area.

When an amplifier is fitted with this type of circuit maximum output may only be obtained when a speaker of the correct impedance is used. With a speaker of low impedance the current output will tend to be limited by this circuit, and the output power will be less. No damage, however, will be done to the amplifier (which is possible as shown if there is no protection circuit).

Instead of using discrete components for an audio output stage an integrated circuit may be used, several circuits being available. I.C. manufacturers never show the internal circuit and the external circuit on the same diagram which makes it difficult to see how the i.c. works since either diagram is useless on its own. A typical i.c. (TBA641) is shown in figure 3.14 giving both internal and external circuits. Much of the complexity of integrated circuits is due to bias and other voltage supply circuits; to simplify the matter the main signal path has been shown in thick lines. The signal is fed through C_1 to pin 7 and the base of Tr_1 which forms a Darlington pair with Tr_2. The load of Tr_2 is Tr_3 which is connected to form a constant-current circuit (*i.e.* high impedance dynamic load). Tr_2 feeds Tr_4 which corresponds to the driver stage and has a load resistor R_1. The circuit is a quasi-complementary one, the output transistor Tr_7 being fed from an emitter-follower Tr_6 (with no phase reversal) while the other output transistor Tr_8 is fed from the common-emitter amplifier (with phase reversal) Tr_5. R_1 and Tr_6 are bootstrapped through pin 12. D.C. feedback from the output is by R_2 to the emitter of Tr_2. A.C. feedback also occurs but is controlled by the values of R_2, R_3 and R_4. C_2 is a d.c. blocking capacitor. Feedback is also provided from the output through the potential divider R_5, C_3 and C_4 to the base of Tr_4. The remaining components in the i.c. are for the application of correct bias. The maximum output power is given as 2·2 watts for 10% distortion at 1 kHz when a 9 volt supply is used.

PERFORMANCE SPECIFICATIONS

These have been described in connection with the f.m. receiver, and similar ones are used in connection with amplifiers. There are two: the DIN specification and the IHF. A brief description of some of the specifications will follow, but a reader wishing to make any of these tests should consult the appropriate specifications.

FREQUENCY RESPONSE

This is fairly simple and is the variation of gain of the amplifier as the frequency is varied. The response is taken by feeding the input with a sine waveform, and varying the frequency and measuring the corresponding output voltage. The measurement is normally taken at a small output level, say 1 watt. If the amplifier has tone controls they should be set to the "flat" position. The amplifier should feed its correct load, a dummy resistive load normally being used. The response is normally plotted in dB against frequency on a logarithmic scale. A good amplifier should have a response within ± 2 dB from, say, 15 to 20,000 Hz. (The DIN specification gives 1·5 dB from 40 to 16,000 Hz). Some manufacturers quote the response as, say, 20 to 20,000 Hz, but this is meaningless unless the variation in gain is given. It might go down 1 dB or 20 dB at 20,000 Hz.

When the frequency response is required of an equalizing amplifier (such as used with magnetic pick-ups) it is better to work with a constant-voltage

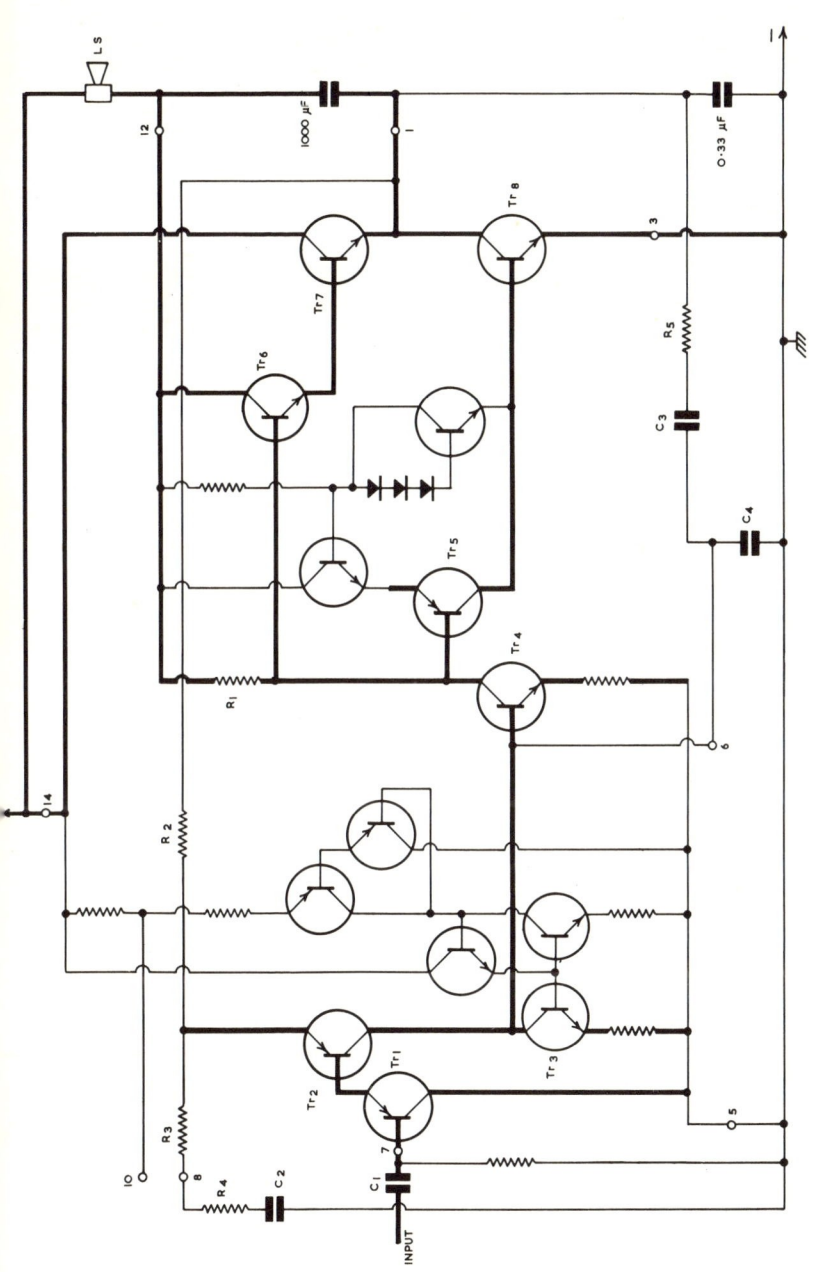

FIG. 3.14 INTEGRATED CIRCUIT POWER AMPLIFIER

output and measure the input voltage. If this is not done it is likely that the amplifier will be overloaded at some frequency.

POWER OUTPUT

One might expect that a single figure would suffice to indicate the maximum power output of an amplifier, but this is not the case. The true power output of any device, provided it is feeding a resistive load, is the product of r.m.s. voltage and r.m.s. current. For tests of power output the loudspeaker is replaced by a resistor of appropriate value, *e.g.* 8 ohms. It must be remembered that with large amplifiers the output is considerable, and a resistor of suitable rating must be used. The power output is normally measured at 1 kHz.

We will first deal with what is called the **continuous sinewave power output**, or **steady-state power** (sometimes called the r.m.s. power). As the name implies, this is the maximum power the amplifier will give under steady-state conditions. To measure it, the input voltage is increased until maximum power is obtained without appreciable distortion, *i.e.* just before clipping occurs. The power output is then V^2/R where V is the r.m.s. output voltage and R is the value of the output load. If a stereo amplifier is being tested it is necessary to state whether both channels are in operation or just one. Due to the additional load on the h.t. supply, when both channels are driven, the h.t. voltage may drop and hence the output per channel is less if both channels are driven. It might be 60 watts with only one channel operating, but only 50 watts per channel when both are in operation. When doing this test the full output should not be maintained for long. The heat sinks may not be designed to deal with this method of operation because the amplifier will not have these conditions when operated normally on music or speech.

The second type of power output is called the **dynamic power output** or **music power output.** Since practically all amplifiers are class-B, the h.t. current consumption is proportional to the output voltage (and current). Unless a stabilized power supply is used the h.t. voltage will therefore drop from no output to full output. However, if the full output is of short duration the h.t. will not drop appreciably, due to the smoothing circuits. Thus, the amplifier will give a greater power for a short period when the h.t. is approximately the same as that on no output, than when the output is maintained and the h.t. voltage has dropped to some lower value, depending on the regulation of the power supply. This is more like the conditions the amplifier will normally operate on since loud passages of music are generally of short duration. The difference between the steady state and dynamic outputs will depend on the design of the amplifier and should be the same if a stabilized power supply is used. For an amplifier without a stabilized power supply the dynamic power may be 1·5 times the steady state of power output. Thus an amplifier which gives a steady state output of 50 watts may give a dynamic power output of 75 watts. In a stereo amplifier the dynamic output is usually quoted as the total output from both channels, *i.e.* an output of 150 watts in the example quoted above.

In class-A amplifiers, since the h.t. consumption is independent of the power output, the dynamic and steady state power outputs should be the same. Also, the heat sinks must be designed to dissipate the full power as this is a maximum under no-load conditions.

A third power rating is sometimes quoted which is most misleading and called "peak power rating". This is obtained by multiplying the peak voltage by the peak current. This is a dishonest way of boosting the power output of an amplifier as peak power has no real meaning in this connection. Since the peak voltage is $\sqrt{2} \times$ the r.m.s. voltage and the peak current is $\sqrt{2} \times$ the r.m.s. current the peak power is twice the normal power output.

Obviously, care must be taken when reading the specifications on the power output of amplifiers if correct comparisons are to be made. It is interesting to note that since November 1974 it has been obligatory in the U.S.A. to give the steady state power output. The maximum power output will depend on the load resistor used which must be stated. If the amplifier is designed for 8 ohms load it will give reduced maximum output on 16 ohms. If it is used with a 4 ohm load the output will be greater, but care must be taken under these conditions as excessive transistor dissipation may take place. The power output will not normally be twice as great (as expected by theory) since, if a limiter protection circuit is used, this will prevent twice the current flowing compared with the correct load of 8 ohms.

POWER BANDWIDTH

As mentioned, the maximum power is normally measured at 1 kHz. The amplifier would not be of much value if the maximum power at other frequencies was much less which can occur at high frequencies. To indicate how'the maximum power output is maintained over the frequency band the term "power bandwidth" was introduced. This is defined as the range of frequencies over which the maximum power output is not less than half of that at 1 kHz. Half power corresponds to -3 dB, and hence it is the range of frequencies over which the power is not more than 3 dB below that at 1 kHz. The power bandwidth should be large and may be 15 Hz to 30kHz in a good amplifier.

DISTORTION: TOTAL HARMONIC DISTORTION (T.H.D.)

One may assume that lack of uniform frequency response is a form of distortion. Under this heading of distortion we shall only consider distortion produced as a result of non-linearity in the amplifier. However good the amplifier there will be some distortion. Thus, if a pure sine waveform is applied to the input of the amplifier the output waveform will not be a pure sinewave, but will show some distortion, which can be represented by harmonics of the fundamental frequency. Although one can, by using a harmonic analyser, determine the magnitude of each harmonic these figures are not usually given. Instead, the total harmonic distortion (T.H.D.) is quoted. This is the ratio of the r.m.s. value of all the harmonics in the output relative to the r.m.s. value of the output and is quoted as a percentage. Alternatively, the ratio may be expressed in dB.

The method of measurement is fairly straightforward, although precautions are necessary. The equipment required is a distortion measuring unit and an audio frequency oscillator having little distortion. The distortion-measuring unit consists of a sharply tuned rejection filter (a notch filter), which removes the fundamental and leaves the harmonics. The total voltage is measured and then the output of the filter is measured. The ratio of the two then gives the total harmonic distortion. If the voltage output is 10 volts and the voltage out of the filter (*i.e.* the harmonics only) is 20 mV, or 0·02 V, the percentage distortion is

$$\frac{0·02}{10} \times 100 = 0·2\%$$

or, expressed in dB (ratio 1/500) $= -54$ dB.

Obviously, the input waveform from the audio frequency oscillator must contain a very low percentage of harmonics (much less than that in the amplifier under test), and so a special low-distortion oscillator must be used, partic-

ularly with hi-fi amplifiers where the T.H.D. will be small. If the output of
the filter is fed to an oscilloscope the harmonics can be seen and often one can
locate the predominant harmonics.

The T.H.D. is normally quoted at maximum power output or the rated
output at 1 kHz. For the fullest information it should be quoted over the
whole frequency range, but this is not usual. When valve amplifiers were in
use it was satisfactory to quote the distortion at maximum output because the
distortion was a maximum under these conditions and always less at lower
outputs. In class-B transistor amplifiers, owing to cross-over distortion the
T.H.D. may be greater at small outputs; hence some manufacturers are quoting
it at low outputs, say 1 watt, for a 50 watt amplifier. The distortion at low
outputs is really more important because (for most of the time) the amplifier
will be operating under these conditions. The period of maximum output,
particularly with classical music, is very small and, in any case, some distortion
under these conditions is not very noticeable.

The less the distortion the better the amplifier, and a good high-fidelity
amplifier should have a distortion of, say, less than 0·5% at full output at 1
kHz, but may be as low as 0·1% or even less. Over the frequency range 40–
15,000 Hz it may be as low as 0·5%.

INTERMODULATION DISTORTION

If two frequencies were fed into a perfect amplifier the output would
consist of only two frequencies, and the amplitude of one would not be
changed by the other. However, owing to non-linearity in the amplifier it is
found that the amplitude of one frequency is varied by the other, *i.e.* one
frequency is modulated by the other. This is called **intermodulation distortion.**
Generally, the test for this is done with a relatively high frequency, say 5 kHz,
together with a low frequency of, say, 100 Hz. The amount of intermodulation
will depend on the relative amplitudes of the two inputs which must be stated.
The distortion may be quoted as the percentage variation of the amplitude of
the higher frequency, and this is used by I.H.F. The modulation process
produces sum-and-difference frequencies, and intermodulation distortion can
be expressed in terms of these. D.I.N. give test frequencies of 250 and 8,000
Hz in the ratio of 4 to 1. The actual amplitudes of the signals must also be
stated. The intermodulation distortion should be less than 1% in a good
amplifier.

There is no direct relationship between intermodulation distortion and
total harmonic distortion, but an amplifier with large total harmonic distortion
is likely to have a large amount of intermodulation distortion.

DAMPING FACTOR

This is really a method of giving the output impedance of the amplifier.
The lower the output impedance the greater will be the damping of the loudspeaker and the less the "ringing". The damping factor is the ratio of the
speaker impedance (the correct value for the amplifier) to the output impedance
of the amplifier. This may have a value of, say, 15 for a 4 ohm load.

The following apply to a complete amplifier, *i.e.* preamplifier and power
amplifier, but are conveniently dealt with here.

SIGNAL-TO-NOISE RATIO

When there is no input to the amplifier there should be no output but, in
practice, there will be some output due to noise generated in the amplifier due
possibly to hum. The magnitude of this output will depend on the setting of
the volume control and on the amplifier input being used. The noise will be

greater for the more sensitive pick-up input than for the tape recorder input. The output may be quoted with the input open circuited or short circuited. Usually it is given with the volume control at maximum and the input short circuited. The noise output is expressed as a ratio to the maximum output and commonly expressed in dB. Thus, if the normal maximum output voltage is 10 volts and the output due to noise is 0·1 volt, the ratio is 10/0·1 or 100/1 or 40 dB. The contribution due to hum can be checked by using a high-pass filter in the output to remove the low frequency of the hum. The signal/noise ratio of a good amplifier is 65 dB for the magnetic pick-up input and, say, 75 dB for the tape recorder input.

In some cases the noise may not be taken as a ratio of the maximum output, but some lower value which must be stated. A weighting network is sometimes used when measuring the noise level, which allows for the annoyance value of the noise.

INPUT SENSITIVITY

This is the signal required on the various inputs to give full output from the amplifier with the volume control at a maximum. For the magnetic pick-up input this may be 2 mV, and for the tape recorder input 200–500 mV. There appears to be no standard for these figures which vary.

INPUT IMPEDANCE

This is the impedance looking into the input socket. For a magnetic pick-up it is usually 47 kΩ, and for the tape recorder input, say, 100 kΩ but this varies. Again there does not appear to be a standard.

CHAPTER 4

PREAMPLIFIERS AND TONE CONTROL CIRCUITS

I N this chapter we shall consider the section which feeds the power amplifier— namely the preamplifier. In a small portable f.m. receiver there will be no preamplifier as the output of the demodulator will normally be large enough to feed the power amplifier direct. A volume control will, of course, be necessary, being simply a variable potentiometer usually connected directly across the output of the demodulator. Loudness compensated volume controls may be used: see end of chapter. The value of this volume control is commonly 5 to 20 kΩ. A simple tone control may be used, such as a capacitor and variable resistor in series, as described in *Radio Servicing*, Volume 2.

In hi-fi amplifiers the preamplifier is much more complex and often switching is incorporated so that the amplifier can be used in conjunction with several pieces of equipment, *e.g.* tuner unit, record player, and tape recorder. The preamplifier may include a relatively complex tone control circuit. A block diagram of a typical arrangement is given in figure 4.1. Four inputs have been shown: phono or record player; tuner unit; auxiliary input; and

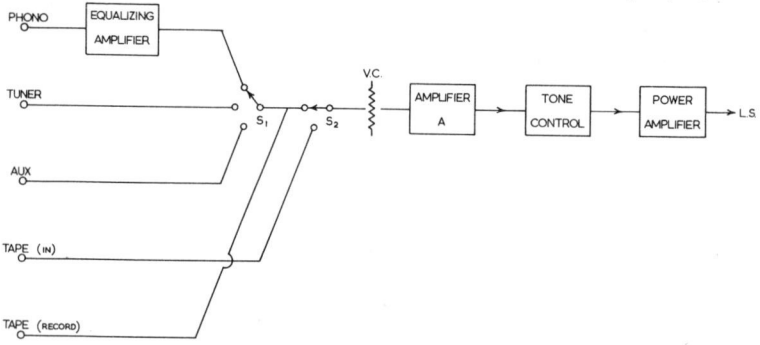

FIG. 4.1. BLOCK DIAGRAM OF HI-FI AMPLIFIER

tape recorder. To correct for the frequency response of the pick-up (now usually magnetic) a special equalizing amplifier is fitted to this input. The amplifier increases the magnitude of the signal in comparison with the other inputs and compensates for the pick-up frequency response. This amplifier will be dealt with in the section on pick-ups in Chapter 10. The switch S_1 selects the output of the equalizing amplifier, the tuner unit or the auxiliary socket (which, for example, can be used for another tape recorder). The output from this switch is taken to the socket marked TAPE (RECORD). This socket is connected to the input of the tape recorder, and hence any signal that is fed into switch S_1 from the three inputs can be recorded, as well as being fed to the preamplifier and main amplifier. When it is required to play the tape recorder then switch S_2 is operated, so connecting the output of the tape recorder to the amplifier. Switch S_2 is used so that the output of the tape recorder is not connected to the input. Switch S_2 is followed by a volume control VC and then a flat amplifier. A tone control stage follows, which feeds the main power amplifier. The sensitivity of the PHONO socket should be about 3 mV with an input impedance of 47 kΩ. The sensitivity of the other three inputs will be 100-250 mV with an input impedance of, say, 50 kΩ, but it may be higher in some equipment. The TAPE (IN) socket will, of course, have the same figures. The TAPE (RECORD) socket will have an output the same as the input to the other sockets as, in this diagram, there is a direct connection.

A more complicated arrangement is shown in figure 4.2. The PHONO input feeds through the equalizing amplifier and switch S_1 (which is closed on

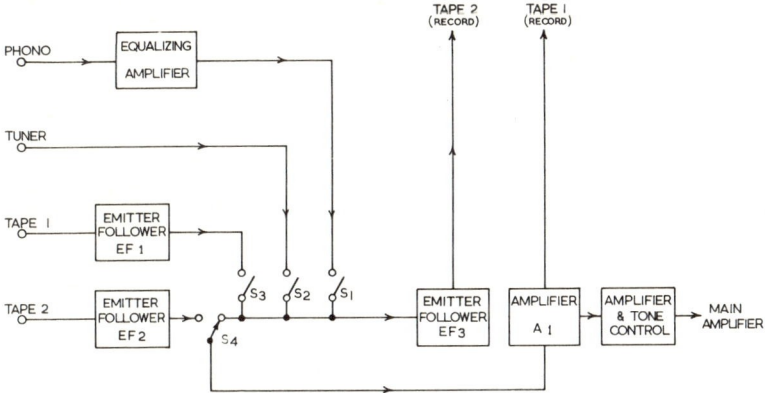

FIG. 4.2. BLOCK DIAGRAM OF HI-FI AMPLIFIER WITH MEANS FOR MONITORING TAPE RECORDING

PHONO) through S_4 to the amplifier A_1 and then to the amplifier and tone control section. The f.m. tuner feeds directly through S_2 (closed on f.m.) and S_4 to A_1 and the amplifier and tone control section. Tape recorder 1 feeds through an emitter-follower EF_1 and S_3 (closed on TAPE 1) and S_4 to A_1 and the amplifier and tone control unit. Any signal from the PHONO or TUNER can be recorded on TAPE 1 from the output from amplifier A_1. In a similar way any signal from the phono, f.m. tuner or tape recorder 1 can be recorded on tape recorder 2 from the emitter-follower EF_3. If it is required to play back from tape recorder 2, then S_4 is operated (it is shown in the non-operating position). The signal from the tape recorder now passes through emitter-follower EF_2 switch S_4 to A_1 and the amplifier and tone control section. However, the circuit is arranged so that if tape recorder 2 is a three-head machine it is possible to record on this machine and at the same time play back through the amplifier, so that the recording can be monitored. It will be seen that with S_4 operated, signals from the phono, tuner or tape recorder 1 are fed (selected by S_1, S_2 and S_3) through emitter-follower EF_3 to the record socket of tape recorder 2. By using emitter-followers in this way the input impedance on the input sockets (excluding the PHONO socket) can be increased (to, say, 1 MΩ) while the output impedance to the record sockets of the tape recorders can be reduced to, say, 5-10 kΩ. Where the equipment is stereo the circuits shown must be duplicated.

Referring to figure 4.1, the amplifier A may be part of the tone control circuit, but if not it is a simple amplifier using resistance load or it may be an emitter-follower. We will now consider the tone control section. Tone controls are usually provided in hi-fi equipment. They may be used to correct for the frequency response of other pieces of equipment, such as the loudspeaker, or just to alter the frequency response to suit a personal preference. Tone controls can vary from very simple circuits, which have already been described in *Radio Servicing*, Volume 2, to complex arrangements. All tone controls make use of elements whose reactance varies with frequency. The inductor and the capacitor are two such elements. The inductor is rarely used because it tends to be both expensive and large, it is difficult to make anything approaching a perfect inductor, and it is apt to pick up hum from stray magnetic fields.

Therefore a capacitor is used almost universally. It is readily available in a
large range of sizes, fairly cheap, and is much nearer the ideal component than
an inductor, *i.e.* its losses are much less. The reactance of a capacitor is given
by $1/2\pi f C$ and hence its reactance decreases as the frequency increases. The
capacitor commonly forms a potential divider circuit with a resistor or resistors,
and first we will look at a few simple circuits. Consider the circuit of figure
4.3. This is a simple potential divider, and the ratio of output voltage V_o to

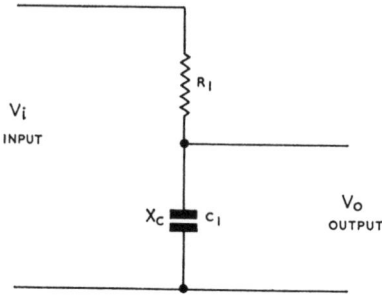

FIG. 4.3. FREQUENCY DEPENDENT POTENTIAL DIVIDER; CIRCUIT TO REDUCE HIGH
FREQUENCIES

input voltage V_i in the general case is given by:

$$\frac{V_o}{V_i} = \frac{\text{Impedance of lower section}}{\text{Total impedance}}$$

In this particular case this becomes:

$$\frac{V_o}{V_i} = \frac{\text{Reactance of } C_1}{\text{Impedance of } R_1 \text{ and } C_1}$$

$$= \frac{X_{C_1}}{\sqrt{R_1{}^2 + X_{C_1}{}^2}}$$

where X_{C_1} is the reactance of $C_1 = \dfrac{1}{2\pi f C_1}$

In order to show response curves we use a horizontal axis of frequency, but a
linear scale is not used as it is difficult to cover the large range of frequencies
involved. Instead a logarithmic scale is used, which is like the scale on a slide-
rule. This is shown in figure 4.4. With this scale we can cover the whole

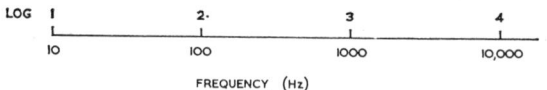

FIG. 4.4. LOGARITHMIC FREQUENCY SCALE

range of frequencies from 10 to 100,000 Hz, or more. In a similar way, since
the ratio of

$$\frac{V_o}{V_i}$$

may be small, the ratio is expressed in dB rather than as a simple ratio or as a percentage. For the present purpose the decibel is defined by:

$$\frac{V_o}{V_i} \text{ in dB} = 20 \log_{10} \frac{V_o}{V_i}$$

Thus the ratio of $\frac{V_o}{V_i}$ is calculated, and the log of this number is obtained from tables. This is then multiplied by 20 to give the dB equivalent of the ratio $\frac{V_o}{V_i}$. It is simpler to look up the dB equivalent on a graph, such as that given in Appendix 2. If V_o is greater than V_i then the ratio is greater than one and the corresponding dB figure is positive. If the ratio of $\frac{V_o}{V_i}$ is less than one then the dB figure is negative. If $V_o = V_i$ the ratio is one and the dB figure is 0. Thus a positive dB figure means an increase in signal (as in an amplifier), whereas a negative dB figure means a decrease in signal (as in a volume control). If a number of circuits are used in cascade, say with gains A_1, A_2 and A_3 expressed as a ratio (the gain figures may be less than one) then the total gain will be given by $A_1 \times A_2 \times A_3$. If the gains are expressed in dB such as B_1, B_2 and B_3 then the overall gain is $B_1 + B_2 + B_3$. If one of these figures is negative then it is, of course, subtracted instead of being added.

A typical response curve for the circuit of figure 4.3 is given in figure 4.5. At low frequencies the reactance of C_1 is high compared with R_1 and the output

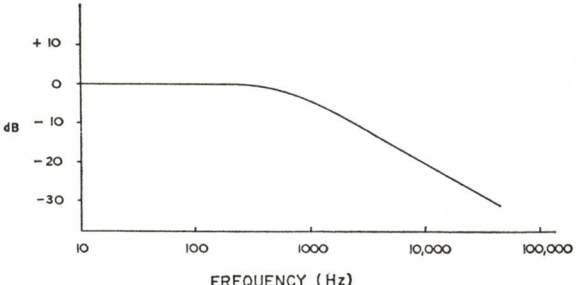

FIG. 4.5. RESPONSE OF THE CIRCUIT OF FIGURE 4.3

is the same as the input, expressed as 0 dB. At a certain frequency the reactance of C_1 becomes comparable with the value of R_1, the output falls and as the frequency increases, this drop in output continues and will drop to a very low value at some high frequency. The frequency at which the output starts to fall is determined by what is called the time-constant of the circuit, which equals the product $C_1 R_1$. The actual values of C_1 and R_1 do not influence the shape of the curve provided the product is the same.

In some cases one does not want the output to fall so much at high frequencies. This can be prevented by using the circuit of figure 4.6, which has the frequency response shown in figure 4.7. As before, at low frequencies the output equals the input (0 dB) since the reactance of C_1 is high compared with R_1. As the frequency is increased the reactance of C_1 decreases and so does the output. However, the resistor R_2 prevents the impedance of the lower section going below the value of R_2 (at very high frequencies the impedance of the lower section will be R_2). Hence the response curve flattens out again as shown.

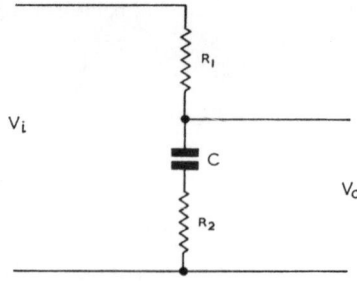

FIG. 4.6.   MODIFIED CIRCUIT TO PREVENT LARGE DROP IN OUTPUT AT HIGH FREQUENCIES

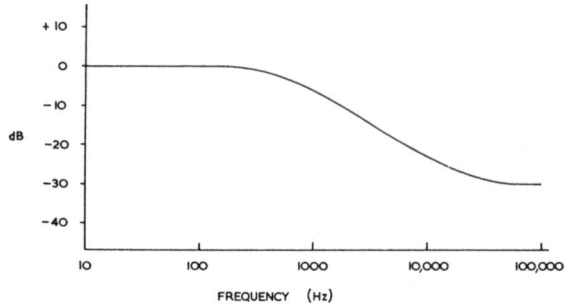

FIG. 4.7.   RESPONSE OF THE CIRCUIT OF FIGURE 4.6

The final figure, at high frequencies, is determined by the value of R_2 relative to R_1.

If the reverse type of response is required the circuit of figure 4.8 is used.

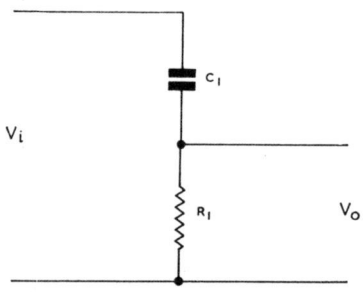

FIG. 4.8.   CIRCUIT TO REDUCE LOW FREQUENCIES

The corresponding frequency response curve is given in figure 4.9. In this case there is no loss in the circuit at high frequencies because the reactance of C_1 will be low compared with R_1. As the frequency is decreased, the reactance of C_1 increases, and when comparable with R_1 the output becomes less than the input as shown in the figure. To prevent the ratio becoming very small at low frequencies this can be limited by the use of a resistor in parallel with C_1, when the response becomes like that shown in figure 4.10.

In order to see the effect of tone control circuits it is useful to have the approximate reactance of the capacitors at different frequencies. A table of reactances is given in Appendix 3, which should prove useful for this and other purposes. Reactance of inductors are also given in Appendix 4.

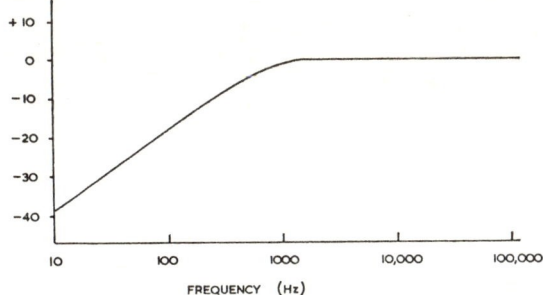

FIG. 4.9 RESPONSE OF THE CIRCUIT OF FIGURE 4.8

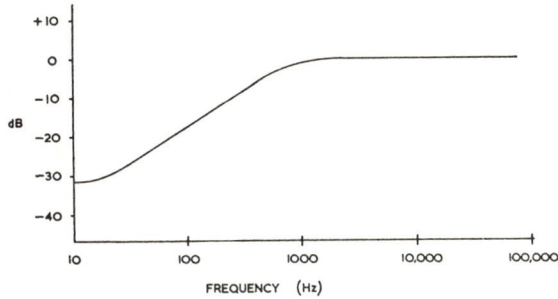

FIG. 4.10 RESPONSE OF THE CIRCUIT OF FIGURE 4.8 BUT WITH A RESISTOR IN SERIES WITH C_1

In a good tone control circuit it is desirable not only to be able to decrease (or cut) the response at one end of the frequency range, but also to be able to increase (or boost) the signal. If such a tone control circuit is used then the gain at middle frequencies must be less than the maximum possible. A common type of circuit is given in figure 4.11, where the left-hand portion of the circuit deals with low frequencies, while the right-hand portion deals with high frequencies (treble or top). We will deal with the two halves separately, although, in practice, there will be some interaction between them. Consider

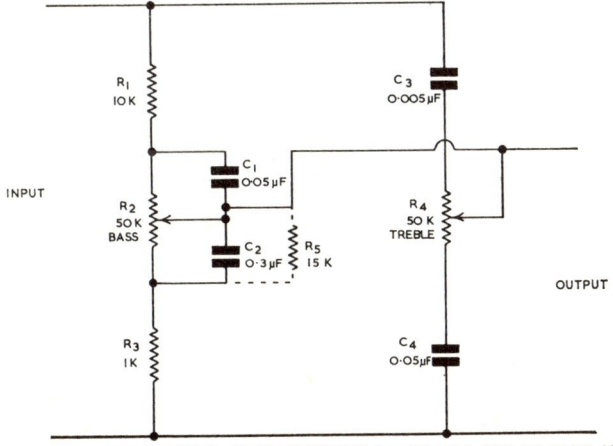

FIG. 4.11 PASSIVE TONE CONTROL CIRCUIT GIVING BOOST AND CUT AT HIGH AND LOW FREQUENCIES

first the right-hand portion. To see the effect of varying R_4 it is best to consider the two extreme cases, one with the slider at the top and the other with the slider at the bottom. Consider first the case of the slider at the top, when the circuit becomes as shown in figure 4.12(a). At low frequencies the reactances of C_3 and C_4 will be fairly high compared with R_4. It should be noted that C_4

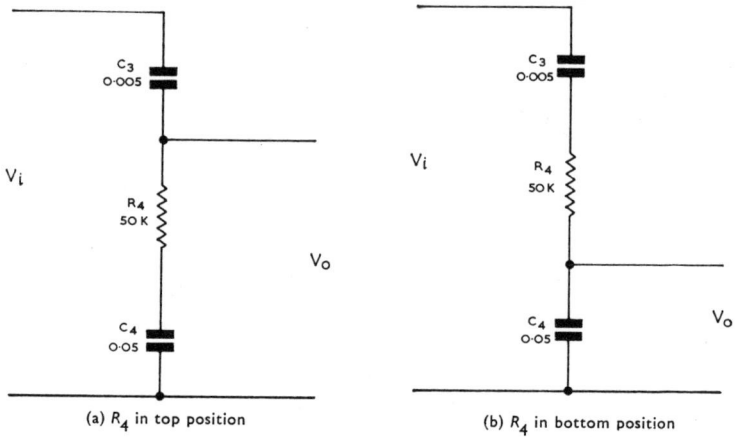

(a) R_4 in top position   (b) R_4 in bottom position

FIG. 4.12. EQUIVALENT CIRCUIT OF RIGHT-HAND PORTION OF FIGURE 4.11

is ten times the capacitance of C_3, and hence will have a tenth of the reactance. For example, at 100 Hz the reactance of C_4 is about 32 kΩ. Thus, neglecting the effect of R_4, the output will be about one-eleventh of the input. As the frequency is raised the reactance of C_4 will eventually become much smaller than the value of R_4, so the lower section approaches the value of R_4. For example, at 10 kHz the reactance of C_4 is only 320 Ω and can be neglected in comparison with R_4. At the same time the reactance of C_3 is decreasing, and at 10 kHz its reactance is about 3·2 kΩ, which is small compared with R_4. Thus, practically all the input appears in the output. The circuit therefore boosts the high frequencies, a typical response curve being given in figure 4.13. This is, of course, the maximum treble boost position of R_4. Next consider the condition when R_4 is at the bottom position. The circuit is now as figure 4.12(b). At low frequencies, when the reactances are high compared with R_4, the circuit behaves as before and the output is about one-eleventh of the input. As before, as the frequency increases the reactances decrease, and at,

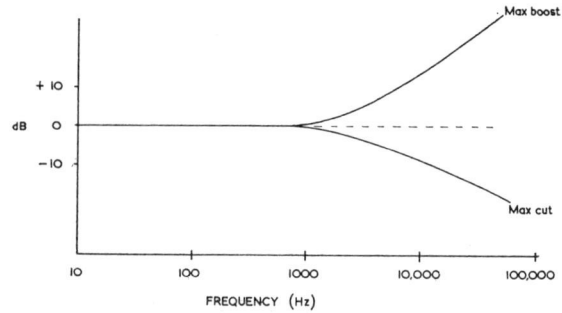

FREQUENCY (Hz)

FIG. 4.13. TYPICAL RESPONSE CURVES OF THE RIGHT-HAND PORTION OF THE CIRCUIT OF FIGURE 4.11

say, 10 kHz the reactance of C_4 is 320 ohms and that of C_3 is 3·2 kΩ. The reactance of C_3 is now small compared with R_4 and can be neglected, hence the circuit consists of a potential divider with a top section of resistance 50 kΩ and a bottom section of reactance 320 Ω. Therefore the output will be much less than the input, the actual ratio being given by

$$\frac{X_{C_4}}{\sqrt{R_4{}^2 + X_{C_4}{}^2}} = \frac{320}{\sqrt{50,000^2 + 320^2}} = 0.0064$$

This is, of course, the position for maximum treble cut, and a typical response curve is given in figure 4.13.

We will now consider the left-hand portion, first with the slider of R_2 in the top position. The corresponding circuit is given in figure 4.14(a). C_1 has been omitted since it is shorted out by R_2. At high frequencies the reactance of C_2 will be low and short out R_2. For example, at 10 kHz the reactance of C_2 is

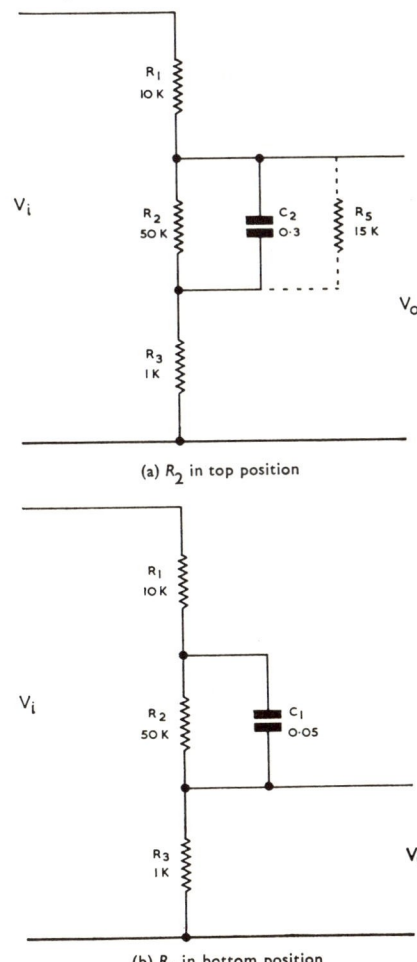

(a) R_2 in top position

(b) R_2 in bottom position

FIG. 4.14. EQUIVALENT CIRCUIT OF THE LEFT-HAND PORTION OF FIGURE 4.11

about 50 ohms, which is small compared with R_1 and R_3. Thus we can consider the potential divider as two resistors R_1 and R_2, and the output will be one-eleventh of the input. At low frequencies the reactance of C_2 will be high, and at 50 Hz it is approximately 10 kΩ. Hence, the impedance of the bottom half of the potential divider is much higher (approximately 10 kΩ) and hence the output is about half the input (difficult to calculate exactly). Thus the output is increased at low frequencies, and this is the maximum bass boost position. A typical response curve is given in figure 4.15. If R_5 is added it prevents the very low frequency response becoming too great, since the impedance of

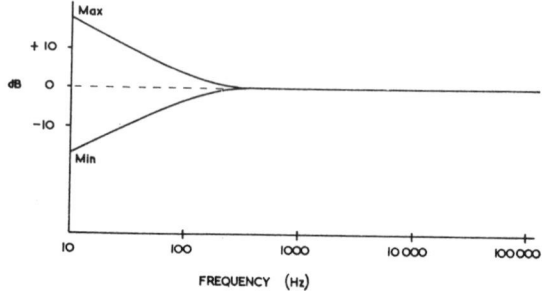

FIG. 4.15. TYPICAL RESPONSE CURVES OF THE LEFT-HAND PORTION OF FIGURE 4.11

R_2, C_2 and R_5 in parallel cannot exceed a value of rather less than 15 kΩ. Consider now the slider of R_2 at the bottom position. The circuit is now as figure 4.14(b). At high frequencies the reactance of C_1 will be fairly low, and at 10 kHz it is approximately 320 ohms, and therefore shorts out R_2. Since this is fairly small compared with R_1 and R_2 the circuit can be considered as a simple resistance potential divider with an output one-eleventh of the input. At low frequencies the reactance of C_1 becomes relatively large, and at 50 Hz its reactance is 64 kΩ. The impedance of the top portion is now much higher (not too easy to calculate) but, say, 50 kΩ, and therefore only a small output will result. This is the maximum bass cut position of R_2 and a typical response is given in figure 4.15.

The explanations of this circuit have been rather simplified, as it has been assumed that the circuit is being fed from an amplifier with zero output impedance, and that the circuit is feeding another amplifier having an infinitively high input impedance. Also, one circuit will react to some extent on the other, as one circuit tends to load the other. There are many possible circuits of this type which are referred to as **passive circuits** because no active elements (such as transistors) are used.

The next circuit is an active circuit since it uses a negative feedback circuit between the input and output of a transistor amplifier. The circuit is shown in figure 4.16. In this simplified circuit no blocking capacitors or bias arrangements are shown. It must be remembered that the frequency response of a negative feedback amplifier is the inverse of the frequency response of the feedback network. For example, the greater the amount of feedback the less the gain. In this circuit R_2 is the treble control and R_5 the bass control. Consider first R_2, and assume the slider is in the top position. The exact analysis of the circuit is difficult and hence only a qualitative description will be given. In this position feedback is through R_1 and C_1. At low frequencies the reactance of C_1 will be high and there will be little negative feedback (n.f.b.), and hence the gain will be relatively high. As the frequency is increased the reactance of C_1 decreases, so increasing the n.f.b. and reducing the gain. This is, therefore, the position of maximum treble cut. Next consider R_2 at the bottom position. There will now be little feedback due to the high value of R_2. As the

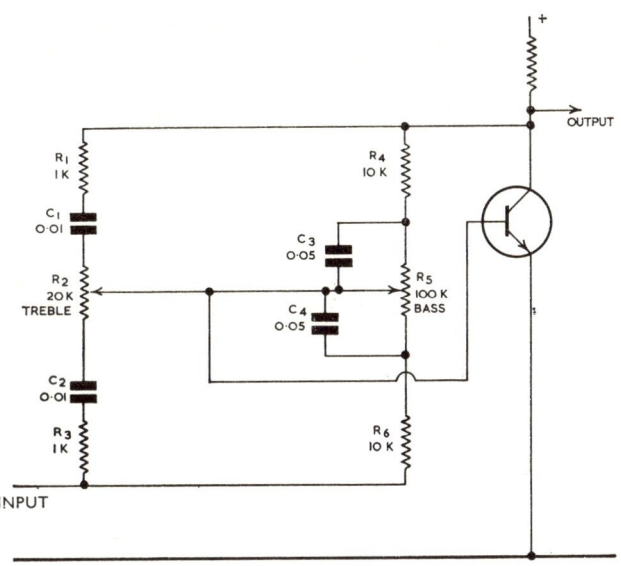

FIG. 4.16. NEGATIVE FEEDBACK TONE CONTROL CIRCUIT

frequency increases the reactance of C_2 will decrease, and hence more of the input signal will be fed to the base and the effective gain increased. This is the position of maximum treble boost. Consider now R_5 and assume the slider is in the top position. C_3 is now shorted out by R_5. At low frequencies the reactance of C_4 will be high, and hence there will be considerable n.f.b. through R_4 which will reduce the gain. As the frequency is increased the react-ance of C_4 decreases and reduces the feedback. This position therefore gives maximum bass cut. With R_5 in the bottom position C_4 is shorted out. At high and middle frequencies the reactance of C_3 will be low, and hence there will be considerable feedback and the gain will be reduced. At low frequencies the reactance of C_3 will increase, and there will be less feedback and the gain will increase. This is then the position of R_5 for maximum bass boost. There are a number of variations of this n.f.b. circuit, but all work on the same general principles.

LOUDNESS COMPENSATION

The frequency response of the human ear is not the same at all loudness levels, and typical curves are given in figure 4.17. These curves indicate the level of sound required to produce the same loudness in the ear. At −10 dB the response is approximately uniform. At a low sound level of −40 dB it is seen that the sound level must be increased at low and high frequencies to hear the same loudness. In other words, the response of the ear drops off at low and high frequencies. If the sound from a loudspeaker is being reproduced at the same level as the original, this change in response is not important. How-ever, if the sound level of reproduction is less than the original, the listener will hear less low and high frequencies than he should do. In general the sound level used in the home is much less than the original. If one runs an amplifier so that the sound level is the same as in the concert hall the neighbours are likely to complain. Obviously, one can overcome this effect by altering the setting of the tone controls when playing at low levels, but it is easier if this is done automatically. Thus we want an amplifier that gives an output or has a

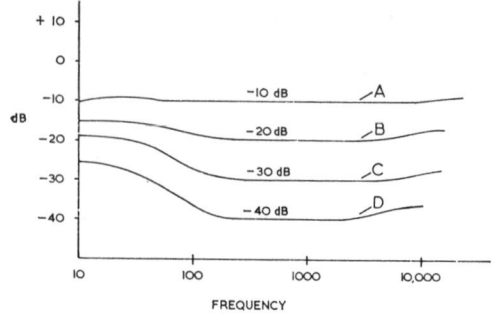

FIG. 4.17. SIGNAL LEVELS TO GIVE EQUAL LOUDNESS IN THE EAR

frequency response like figure 4.17 and varies with the sound volume. This is fairly easy to approximate to by putting a compensating circuit across part of the volume control, and a simple circuit is shown in figure 4.18. In this case C_1 and R_1 are connected from a fixed tap on the volume control to the earthy end. When the slider is above the tap, C_1 and R_1 will have little effect; but

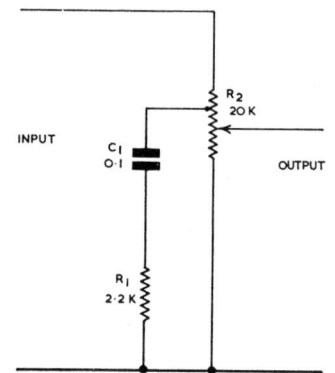

FIG. 4.18. LOUDNESS CONTROL CIRCUIT

below the tap (at low volume settings) the response will be reduced at middle and high frequencies because of the reduced reactance of C_1 (1·6 kΩ at 1 kHz), and so the low frequencies will be boosted relative to the output at other frequencies. There is no compensation for boost at the high frequency end in this simple circuit. It is important to note that this compensation only comes into action when the volume control is set to a low level and does not operate if, for example, there is a quiet piece of music. The position of the tap must, of course, be related to the actual sound volume from the loudspeaker. A more complex circuit is given in figure 4.19, where again there is a tap on the volume control. When the slider is above the tap the compensating circuits will have little effect until near to the tap. When below the tap the high frequencies will be boosted by R_1 and C_1, the reactance of C_1 at 10 kHz being approximately 5 kΩ. At middle and high frequencies the reactance of C_2 will be low (the reactance of C_2 at 1 kHz is about 40 Ω) and short out R_2 so that the bottom section of the volume control has a parallel circuit of fairly low impedance which will reduce the output. At low frequencies the reactance of C_2 will increase, so increasing the impedance of this parallel circuit and increasing the output. Thus there will be boost at both high and low frequencies. In some amplifiers this loudness compensating circuit can be switched out.

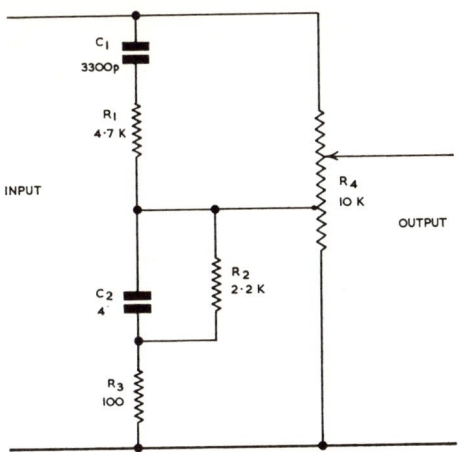

FIG. 4.19. MORE COMPLEX LOUDNESS CONTROL CIRCUIT

HIGH FREQUENCY AND LOW FREQUENCY FILTERS

Some amplifiers are fitted with switches that bring in high frequency and low frequency filters. The low frequency filter reduces the response at very low frequencies and is mainly used to reduce rumble from a turntable. This may simply reduce the value of one of the coupling capacitors in the amplifier or may be more elaborate. The high frequency filter cuts the response at very high frequencies and is used to reduce the noise. This commonly switches in a capacitor in parallel with the load resistor of the preamplifier, but can be a more elaborate filter.

POWER SUPPLIES

As normal power supplies were described in *Radio Servicing*, Volume 2 they will not be considered in this volume. A simple stabilizer circuit was described and in this chapter we will deal with more complex voltage stabilizers. It might first be mentioned that some receivers use a simple stabilizer circuit on critical stages, *e.g*, the oscillator of the v.h.f. tuner. A simple circuit is given in figure 5.1, which makes use of a voltage dependent resistor (VDR) R_2. This may be replaced by a zener diode. If the input voltage

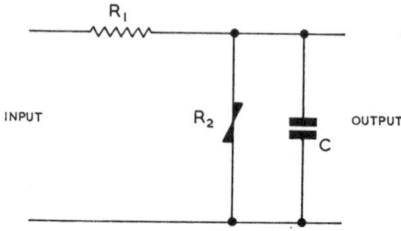

FIG. 5.1.   USE OF VDR IN A SIMPLE VOLTAGE STABILIZER CIRCUIT

rises there will be smaller percentage change of voltage across R_2 than across the input, because of the constant voltage characteristic of a VDR or zener diode. Capacitor C is added to form a low reactance path for a.c. This circuit may be used on battery or mains receivers.

Since the output stages of amplifiers nearly always operate under class-B conditions they present a variable load to the power supply; if an unstabilized supply is used there will be corresponding variations of voltage because of the relatively poor regulation of a rectifier supply. Many receivers and amplifiers do not use a stabilized power supply, but there are receivers of this type which may use a simple stabilizer on some of the v.h.f. stages. Basically, one may use a stabilizer to maintain a constant voltage on all sections of the receiver; or a stabilizer feeding all but the power amplifier, since the current and voltage will be less, results in a much cheaper stabilizer. It maintains a constant voltage on all stages except the output stage, and hence prevents distortion on early stages or oscillator drift due to changing voltage.

Also, it prevents variations in supply voltage being fed back to earlier stages and causing oscillation or "motor boating". Obviously, the output stages are not working under the best conditions because the supply voltage will be a minimum when maximum output is required. The type of stabilizer is basically the same whatever arrangement is used, the only difference being the current and voltage output. A typical stabilizer circuit is shown in figure 5.2. The input voltage is -56 V and the output -15 V, which is a greater difference than one might expect. These voltages are the kind which would occur where the stabilizer was not feeding the output stages. The -56 V (unstabilized) would be for the output stages and the -15 V for the remainder of the equipment. R_6 and C_1 form a simple filter to remove some ripple, and the series control transistor is Tr_3. The control amplifier is a long-tailed pair consisting of Tr_1 and Tr_2. Tr_2 base is fed with a constant voltage from across the zener diode D_1, which is supplied through R_1, from the output voltage. Tr_1 base is fed with a fraction of the output voltage from the potential divider formed by R_7 R_5. It has already been shown that in a long-tailed pair the two bases must be at approximately the same voltage if the transistors are conducting. Thus the output voltage sets itself to such a value that the voltage on the

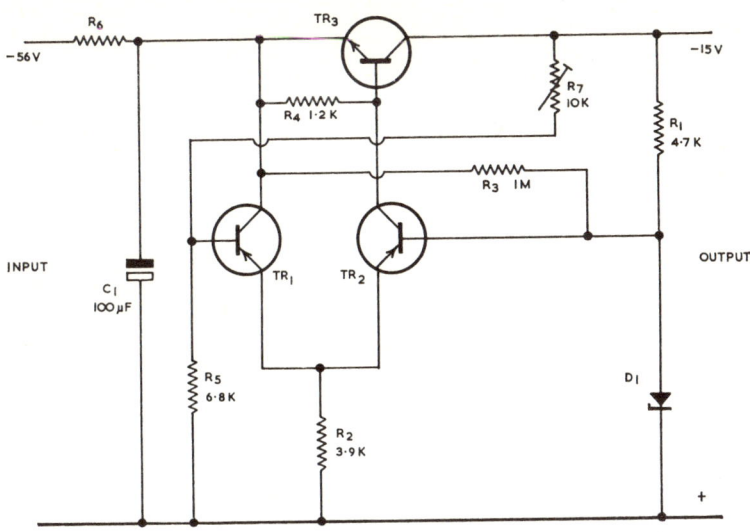

FIG. 5.2. VOLTAGE STABILIZER CIRCUIT

base of Tr_1 is approximately that on the base of Tr_2. The collector current of Tr_2 flows into the base of Tr_3. Resistor R_4 is added to carry any collector-base leakage current of Tr_3. Suppose that the output voltage rises. This will cause the base voltage of Tr_1 to rise, and hence Tr_1 passes an increased current. This will cause an increased voltage across R_2 and since the base of Tr_2 is at a fixed voltage, reduces the current of Tr_2. The reduced base current of Tr_3 will cause the voltage across it to increase and so reduce the output voltage until equilibrium conditions are reached. R_3 is added to start the circuit. Without this resistor there would be no current in Tr_2 on first switching on, since there would be no voltage across D_1 because there would be no output voltage. No current in Tr_2 means no current in Tr_3, and therefore no output. With R_3 present a small current will flow from the input so that a voltage is produced across D_1, this causing Tr_2 to conduct and bring the circuit into normal operation. The current output from such a circuit is mainly determined by the allowable dissipation of Tr_3.

Another circuit is given in figure 5.3. In this case the control transistor is connected as an emitter-follower and not as in figure 5.2. This circuit would be suitable for supplying a whole amplifier. Tr_1 is the main series control transistor, and is fed from the current amplifier Tr_2. The emitter of Tr_3 is supplied with a constant voltage from across the zener diode D_1, which is fed from the output by R_3. The base of Tr_3 is fed with a fraction of the output voltage by the potential divider R_4, R_5 and R_6; R_5 is preset so that the voltage can be set to the required value. C_2 forms a low impedance path across D_1 and removes any supply ripple. C_3 is used to feed any a.c. component on the output direct to the base of Tr_3 and not reduced in magnitude by the potential divider, as would occur if it were omitted. In this way the stabilization is improved as regards ripple because the loop gain is increased. The load for Tr_3 is R_7 and R_1 (as regards d.c.), C_1 again acting as a smoothing circuit with R_1. Tr_2 is connected as an emitter-follower so that the emitter follows the collector voltage of Tr_3. The emitter current mainly flows out of the base of Tr_1. The emitter of Tr_1 follows the emitter voltage of Tr_2, hence the output voltage is approximately the collector voltage of Tr_3. R_2 completes a path for any collector-base leakage current of Tr_1. If the output voltage rises then the base-

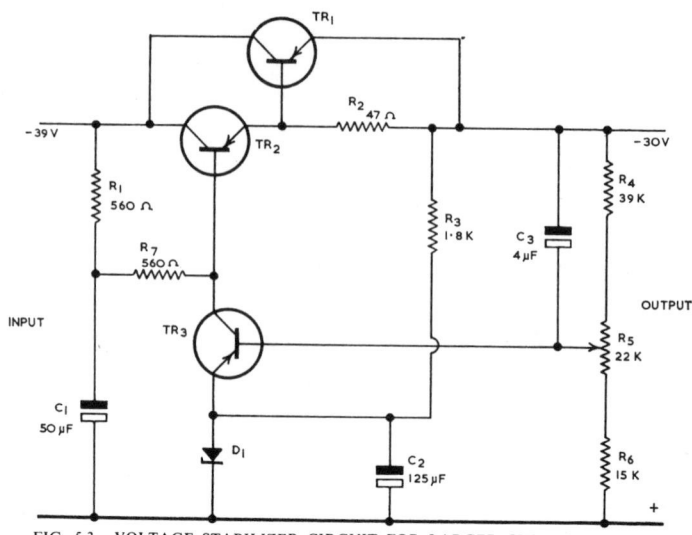

FIG. 5.3.   VOLTAGE STABILIZER CIRCUIT FOR LARGER CURRENT OUTPUT

emitter voltage of Tr_3 is increased. This increases its collector current and results in reduction of collector voltage. This causes a reduction in voltage on the emitter of Tr_2, and on the emitter of Tr_1, so reducing the output voltage until equilibrium conditions are reached. The circuits described have no protection against overload and generally no protection is provided. However, overload or constant current circuits can be added, so that excessive current cannot flow. Thus, if an attempt is made to take excessive current the voltage will drop. A simplified circuit showing how overload protection and current limiting can be done is given in figure 5.4. Tr_1 is the main series transistor (probably fed from a current amplifier) and Tr_2 is the control transistor. The emitter of Tr_2 is maintained at a constant voltage from across the zener diode D_1, fed through R_2. The base of Tr_2 is fed with a fraction of the output voltage from the preset potentiometer R_3. Thus, if the output voltage rises, the base-emitter voltage of Tr_2 is reduced so reducing the current of Tr_2. This,

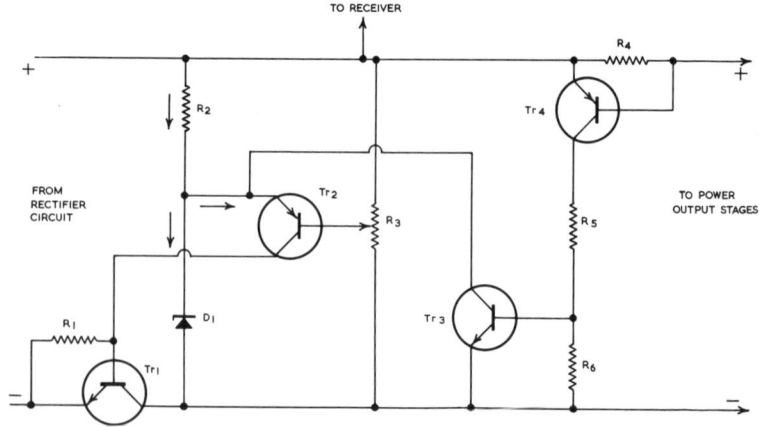

FIG. 5.4.   VOLTAGE STABILIZER CIRCUIT WITH OVERLOAD PROTECTION

in turn, reduces the base current of Tr_1 and so increases the drop across it to reduce the output voltage until equilibrium is reached. The current through R_2 is divided between the zener diode and transistor Tr_2 as shown. The current in R_2 will be approximately constant for a given output voltage. As the load current increases, the current in Tr_2 must increase in order to supply sufficient base current to Tr_1. At some output current Tr_2 will take all the current flowing through R_2 and there will be nothing left for the zener diode. Accordingly, the zener diode voltage will drop and, since this is the reference voltage, the output voltage will also drop. In this way excessive current flow is prevented. If it is required to prevent excessive current flowing in the output stages due to, say, too low a load, the circuit shown on the right-hand side may be used. The current for the output stages flows through a very low resistance R_4 and provided the voltage across this resistor is less than about 0·5 volt, Tr_4 will not be conducting and the circuit operates normally. However, if excessive current flows this voltage will rise and cause Tr_4 to conduct, which also causes Tr_3 to conduct. Tr_3 shorts out the zener diode D_1, and so the voltage is reduced until the cause of excess current is removed. When overload protection is added to the stabilizer the rating of the series transistor usually must be increased, as it must withstand full supply voltage (with zero output voltage) and its dissipation under overload conditions can be high. There are many possible stabilizer and protection circuits and these are only examples.

MAINS-BATTERY RECEIVERS

Many receivers are now made so that they can be operated from mains or batteries. When operated on mains a small power supply is required giving the same voltage as the battery. A normal power supply circuit may be used or a simple stabilizer circuit such as the emitter-follower circuit shown in *Radio Servicing*, Volume 2. Some means must be used to change over from battery to mains. One method is to use a simple change-over switch, it being important that the battery and power supply are not connected in parallel. Another is to have a change-over switch automatically operated when the mains lead is plugged into the receiver. An alternative is to use a stabilized power supply and connect the battery through a diode. This is shown in figure 5.5, where Tr_1 is the series transistor of the stabilizer and supplies a voltage rather greater than the battery voltage. D_1 is the diode in series with

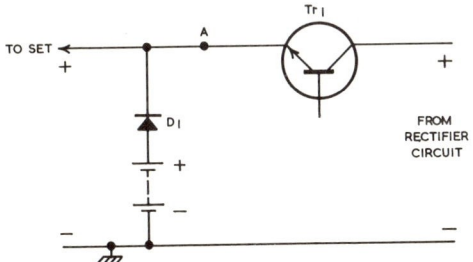

FIG. 5.5. CONNECTION OF BATTERY TO MAINS/BATTERY RECEIVER

the battery. When the mains is connected the voltage at point A will be higher than the battery voltage, but no current will flow into the battery because D_1 is reverse biased. When the mains is disconnected the set will be supplied by the battery through D_1. No current will flow in Tr_1 because it is reverse biased.

CHAPTER 6

F.M. AERIALS

THE kind of receiving aerial to be used depends on the carrier frequency and not on the type of modulation. The chapter title might therefore be misleading. At the relatively high frequency corresponding to v.h.f. it is common to use resonant aerials, *i.e.* the aerial is of such a length that its natural resonant frequency corresponds to the frequency or band of frequencies to be received. The simplest aerial is the dipole, and is shown in figure 6.1. This consists of a single rod split in the centre, the receiver being connected

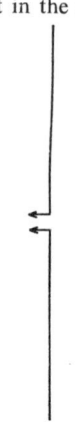

FIG. 6.1. DIPOLE AERIAL

between the two sections. The total length for correct resonant frequency is rather less than half the wavelength. The relationship between frequency and wavelength is given by

$$\lambda = \frac{300 \times 10^6}{f}$$

where f is the frequency in Hz and λ is the wavelength in metres.

Thus, for a frequency of 100 MHz, the wavelength is

$$\frac{300 \times 10^6}{100 \times 10^6} = 3 \text{ metres}$$

The dipole will, therefore, be about 5% less than 1·5 metres. The impedance of a simple dipole is about 70 Ω, and matches the normal 75 Ω coaxial cable.

All v.h.f. f.m. transmissions in this country use horizontal polarization, *i.e.* the electric field of the electromagnetic radiation is horizontal. The dipole should be horizontal and at right angles to the direction of the transmitter. A drawback of the dipole is that it receives equally well in the opposite direction and may pick up interference. It is better to use an aerial that is much more directional because (a) there is a larger induced voltage in the aerial, so improving the signal/noise ratio (particularly important on stereo); (b) less interference is picked up; and (c) there will be less distortion due to multipath reception (*i.e.* the signal is received from more than one direction).

The first step is to add a reflector as shown in figure 6.2, which reduces the pick up from the rear (*i.e.* in the direction of the reflector). The reflector is normally about half a wavelength long and spaced one-quarter of a wavelength

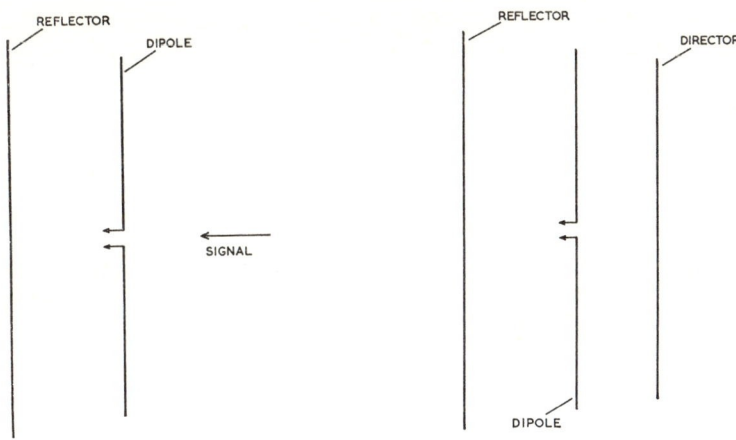

FIG. 6.2. DIPOLE AERIAL WITH
REFLECTOR

FIG. 6.3. DIPOLE AERIAL WITH
REFLECTOR AND DIRECTOR

from the dipole.   The aerial can be further improved by adding what is called
a "director", as in figure 6.3.   More than one director may be used with
advantage. The directors are normally rather less than a half wavelength in
length, and spaced from 0·1 to 0·25 of a wavelength away from the dipole
and each other. The spacing and length are related and are important or the
director may act as a reflector. This type of aerial is often called a "Yagi"
aerial, and is a considerable improvement on the simple dipole because the
voltage it picks up is greater and because it is highly directional, and picking
up only those signals coming in the required direction.

When a reflector and directors are added in this way the impedance of the
aerial is reduced and no longer matches the 75 Ω coaxial cable. To overcome
this a folded dipole is commonly used, as shown in figure 6.4. This has the
effect of increasing the impedance by an amount depending on the ratio of

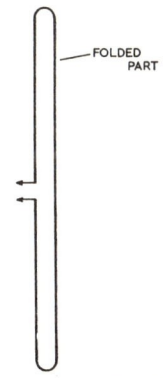

FIG. 6.4.   FOLDED DIPOLE

diameter of the folded part of the aerial to the other part. Usually the two
parts are made the same diameter. In this country a coaxial cable is normally
used between the aerial and the set, but technically this is not correct because
the aerial is symmetrical, or balanced to earth, and the cable is not. An alternative is to use a twin lead, but this does not match the aerial so well because its

impedance is about 300 ohms. Some receivers are made to be fed with a twin lead but can easily be fed with a coaxial cable. A typical input circuit is given in figure 6.5. The aerial is coupled to the first tuned circuit of the receiver by the centre-tapped coil. This is sometimes known as a "balun", since it

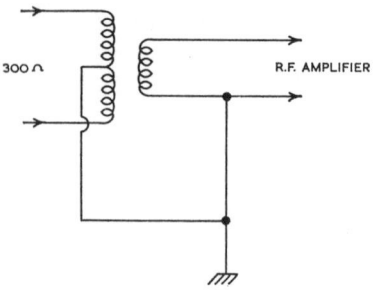

FIG. 6.5. INPUT TRANSFORMER OF RECEIVER

converts the balanced twin lead input to the unbalanced first tuned circuit. The turns-ratio must be chosen so that the cable is matched to the first tuned circuit. A coaxial cable should be connected to one outer lead and the centre-tap (earthy lead). The impedance across these two points will be a quarter of that between the outers and is 75 Ω. It is a quarter because this transformer ratio is 2/1, and the impedance change is proportional to the turns-ratio squared.

The aerial, although it is resonant, must cover the band of frequencies used, i.e. 88 to 98 MHz. The bandwidth of the aerial is related to the diameter of the elements, but if the elements are of a diameter sufficient for mechanical strength the bandwidth is then adequate.

Such an aerial cannot be used on small portable receivers and a telescopic aerial is usual, which is not resonant. Its pick up is poor but adequate in good signal-strength areas but it may pick up interference. Large changes in signal strength may occur with quite small movements of the aerial. Unfortunately, the position for best reception may not correspond with a suitable place for the receiver. Some portable receivers are fitted with an aerial socket so that an external aerial may be used for better results or when using the set in a car.

CHAPTER 7

LOUDSPEAKERS AND HEADPHONES

THE moving-coil loudspeaker is described in *Radio Servicing*, Volume 2 and a knowledge of such a speaker will be assumed. The only other type sometimes used is the electrostatic loudspeaker. This consists basically of two plates about 1 metre square, close together with one plate able to vibrate. By applying a variable voltage to the plates the electrostatic force changes and one plate therefore moves and produces sound waves. It is essential to have a high polarizing voltage. A high audio frequency voltage is also required, which is obtained from a step-up transformer. No more will be said about this rather special speaker.

For efficient operation, a moving-coil speaker must be mounted on a baffle or in a cabinet. If it is not then, at low frequencies, it will produce little output because the radiation from the front will be cancelled by the radiation from the back. This occurs mainly at low frequencies because there is time for the air to move from the back to the front and *vice versa*. This can be overcome by fitting the speaker to a baffle which is a strong board with a hole in it in which the speaker is mounted. If the distance from front to back is long compared with the wavelength of the sound, cancellation does not occur; there is now not time for the air to travel around the baffle. However, for the reproduction of low frequencies the baffle must be large. For 100 Hz it means a baffle about 1 metre across. A baffle is not generally a convenient way of mounting a speaker, and some type of cabinet is normally used.

Before dealing with the cabinet design it is necessary to explain what is meant by "resonance". The cone of the speaker has mass and it is confined to its centre position by the inner and outer supports. At some frequency, like all mechanical systems, the cone will resonate in the same way as an electrical circuit containing inductance and capacitance. Ideally this resonant frequency should be outside the audible range, but this is not practicable. It is, however, made as low as possible and is normally between 50 and 150 Hz. At the resonant frequency the sound output of the speaker is increased and can cause boomy reproduction. The increase depends on the damping or Q factor, (magnification factor) of the loudspeaker. The greater the damping the less the increase. Some damping is produced by the cone supports, some by the cabinet and some electrically. When the cone and moving coil move an e.m.f. is induced in the coil which will cause a current to flow in it and in the output circuit of the amplifier. The force produced by the current, by Lenz's law, is such as to tend to stop the movement or damp the speaker. The magnitude of the current flowing (and hence the damping) will depend on the output impedance of the amplifier which should be low. Modern amplifiers have a low output impedance and the electrical damping is good.

One method of mounting a speaker is to place it in a closed case, which must be airtight apart from a small hole to allow the pressure to equalize between inside and outside as the temperature and barometric pressure vary. This type of enclosure is called the "infinite baffle type". As the cone moves it will increase and decrease the pressure in the cabinet and it moves against the "springiness" of the air which adds to the effect of the cone supports and increases the resonant frequency. For good performance, a speaker used in this type of enclosure should have a low resonant frequency in free air. The smaller the cabinet the greater will be this effect, and hence a large cabinet is desirable, but conflicts with the common requirement of a small speaker for domestic use. To reduce any resonances set up in the cabinet it should be filled with sound-absorbing material such as resin-bonded fibreglass or acetate fibres. To prevent the sides of the box resonating it should be of solid construction, preferably made of dense wood or chipboard. The infinite baffle enclosure is widely used commercially.

Another method of mounting the speaker is the labyrinth enclosure, shown in figure 7.1. Instead of trying to absorb the radiation from the back of the cone the idea of the labyrinth cabinet is to make use of the back radiation

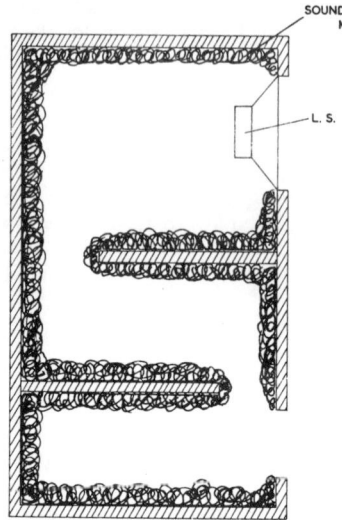

FIG. 7.1. LABYRINTH-TYPE SPEAKER CABINET

and add it to that from the front radiation, at low frequencies. The sound waves radiated from the back are 180° out of phase with those from the front; but, by allowing the back radiation to travel a suitable distance it can be made to be in phase with the radiation from the front. The back radiation travels along the labyrinth and comes out of a hole in the front of the cabinet as shown in figure 7.1. Careful design is required to obtain good results. To prevent resonances at high frequencies the back radiation should be absorbed by suitable sound-absorbing material lining the whole of the inside of the cabinet.

The reflex type of enclosure is shown in figure 7.2. The back radiation at low frequencies is also utilized, but the cabinet space now becomes a resonator, the output coming from a tube or hole at the bottom of the cabinet. The cabinet should be designed to match the speaker and the design is rather critical for best results. The hole at the bottom may be replaced by a passive element sometimes called an A.R.U. (acoustic resistance unit), *i.e.* a diaphragm which is vibrated by the air in the cabinet and will give sound waves from the front which, if correctly designed, will be in phase with those from the front of the main speaker. As before, the loudspeaker cabinet should be lined with sound-absorbent material to prevent resonances occurring at middle and high frequencies.

It is possible to load a loudspeaker with a horn, which can be very efficient. However, if the speaker is to operate at low frequencies the horn must be long (say 10 to 20 feet) and have a large mouth. The size can be reduced by folding the horn. Horns are rarely used in domestic speakers, but they are, of course, commonly used for public address purposes, where good low frequency response is neither required nor desirable.

A loudspeaker is a complex device: the coil does not simply drive the cone back and forth like a piston. This occurs at low frequencies, but not at high

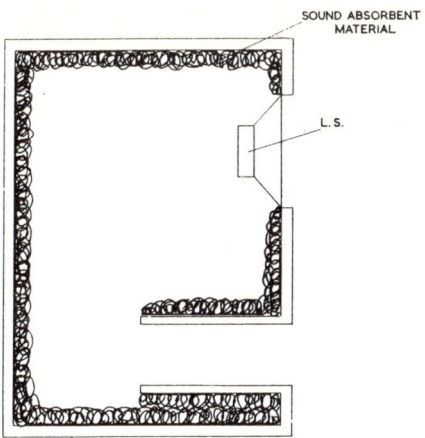

FIG. 7.2. REFLEX-TYPE SPEAKER CABINET

frequencies owing to the fact that the mass of the cone is too large. The cone does not move as one piece but the vibrations break up and cause parts of the cone to vibrate. In practice a speaker is not, of course, dealing with a single frequency but with a number of frequencies at the same time. If a single speaker is used problems arise under these conditions and the high frequencies may be modulated by the low frequencies, etc. To reduce these effects, modern speakers use two or more units. When two are used one deals with low and middle frequencies (say up to 2-5 kHz) while the other, commonly called a "tweeter", deals with the high frequencies. When three units are used one deals with low frequencies up to say 500 Hz (sometimes referred to as a "whoofer" or "bass unit"); the second one deals with the middle range of frequencies (say 500 Hz to 5 kHz called the mid range unit); and the third unit, a "tweeter", deals with frequencies above 5 kHz. Each unit can thus be designed to deal with a limited range of frequencies which is much easier than trying to design a single unit to cover all the audio range. As each speaker covers a limited frequency range there is less chance of one frequency modulating another. The bass and mid range units are made in the same basic way as described in Volume 2, but the bass unit has a larger cone. In some cases the cone of the bass unit is made from expanded polystyrene, which combines rigidity with lightness in weight, i.e. the mass is small. The tweeter may be made like a normal speaker, but smaller; or it may be made differently, one method being shown in figure 7.3. The magnet system M is the same as that of a normal moving-coil speaker, but may be rather smaller. The moving coil is now attached to a diaphragm D which is made conical for strength and is usually of plastic. The diaphragm and coil are held in place radially by a corrugated portion C attached to suitable supports R. The corrugated portion allows movement in an axial direction but not radially, the actual diaphragm movement being small at the high frequencies concerned. Sound-absorbent material may be placed at the back of the diaphragm to reduce any resonances. The diaphragm (cone portion) is usually 3 to 4 cm diameter.

When a number of speakers are used in this way they cannot just be connected in parallel. The tweeter would be damaged by the low frequency signals as it is not designed for large coil movement as is required at low frequencies. If high frequencies are fed to the other speakers they would try to radiate the high frequencies and cause interference with the tweeter and also might have bad resonances at the high frequencies. Accordingly, the units should be fed with their

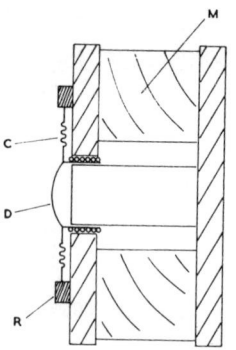

FIG. 7.3. TWEETER LOUDSPEAKER

appropriate range of frequencies by means of a "crossover unit". The basic idea for two speakers is given in figure 7.4. The high frequency (HF) speaker is fed through a capacitor which will have a high reactance at low frequencies.

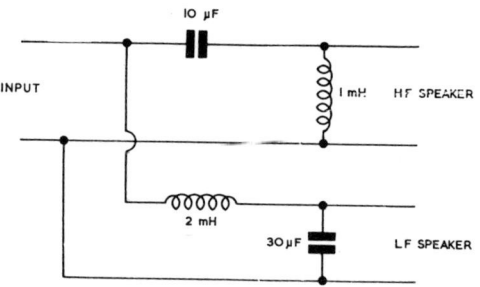

FIG. 7.4. CROSSOVER NETWORK FOR TWO SPEAKERS

Similarly, the inductor will have a low reactance at low frequencies and short them away from the tweeter or HF unit. The series inductor of the LF unit will prevent the passage of high frequencies and the capacitor will py-pass any from the speaker. A possible circuit for use with three speakers is given in figure 7.5. The high frequencies will be developed across the 1 mH inductor and the low frequencies across the 30 μF capacitor. The medium frequency unit is fed from across the parallel tuned circuit, which must have suitable damping so that it covers the appropriate frequency range. Much more

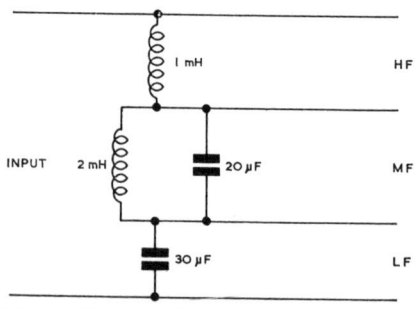

FIG. 7.5. A POSSIBLE CROSSOVER NETWORK FOR THREE SPEAKERS

complex cross-over networks are used in some speakers. The design of these cross-over units, or filters, is difficult because they have to deal with power and feed the low impedance of the speakers. Also, the impedance of a speaker varies greatly with frequency owing to resonances, etc. The use of two or three amplifiers with filters before the amplifiers has been used and at least one commercial unit is available using this arrangement. This makes the filter design much easier.

One method of reducing the resonance in the low frequency unit is to use a servo control; this is called "Motional Feedback" by Philips. In this unit a piezoelectric unit (see pick-ups Chapter 10) is attached to the cone and this will detect the movement of the cone and produce a voltage proportional to its movement. This voltage is fed back to a comparator before the power amplifier (a separate power amplifier is used for the bass unit) so that if the cone does not move as it should a correcting signal will be produced by the amplifier. It is therefore a negative feedback system but includes the speaker in the feedback loop.

There are very many modifications and speaker designs which cannot be considered in this book. In general the larger the volume of the cabinet and the more costly the speaker the better is the result.

HEADPHONES

Hi-fi headphones can give extremely good results, sometimes better than those produced by a good loudspeaker. They have the inestimable advantage of personal listening: others in the room are not annoyed; they can be operated at any volume level without upsetting the neighbours; and they cut out much of the external noise. It is not intended to go into detail of the construction. Most are moving coil or electrostatic. The moving coil ones are similar to a small moving-coil speaker. The electrostatic ones are similar to an electrostatic speaker (or an electrostatic microphone: see Chapter 8). They require a polarizing voltage which is generally obtained from a mains unit. Hi-fi headphones can be expensive, but the better ones give superlative quality. Of course, not everybody likes the idea of wearing headphones. When listening to stereo the image that is produced by headphones is different and disliked by some people.

MICROPHONES

A microphone is the reverse of a loudspeaker. A loudspeaker is required to convert electrical power into acoustic power, whereas a microphone is required to convert acoustic power into electrical power. There are a number of types of microphone depending on their method of operation. They may be listed as follows:

(1) Carbon   (4) Crystal
(2) Moving coil or dynamic   (5) Electrostatic (or capacitor)
(3) Ribbon

The carbon microphone consists of carbon granules between two carbon plates, one of the plates being attached to the diaphragm. The effect of the sound waves is to vary the pressure on the granules which varies the resistance. By applying a voltage to the two carbon plates the current varies according to the sound pressure. The frequency response of this type of microphone is poor, and is not used for high-quality reproduction. It is used in telephones, where the poor response is not a disadvantage and may be an advantage. It will not be discussed further.

The moving-coil microphone is constructed basically in the same way as a moving-coil loudspeaker. In fact, a loudspeaker will act as a microphone, but not of particularly good quality. This type of microphone is now more commonly called a "dynamic" microphone, and its basic construction is given in figure 8.1. It consists of a circular magnet system M with a circular gap like a moving-coil loudspeaker. Attached to the

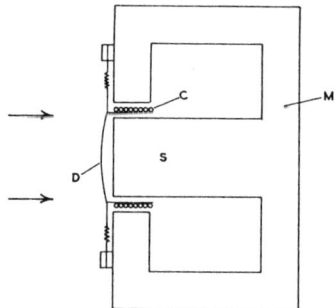

FIG. 8.1 CONSTRUCTION OF MOVING-COIL OR DYNAMIC MICROPHONE

light diaphragm D is the moving coil C, which moves in the magnetic field of the magnet system. To hold the diaphragm in place and prevent it moving radially is a corrugated section, as shown, fixed to a suitable circular support. This allows reasonably free movement of the coil in the axial direction. Thus, when sound waves impinge on the diaphragm it causes the coil to move. As this moves in the magnetic field an e.m.f. is generated. If the microphone is to be sensitive, and also to have a good frequency response, care is necessary in the design and manufacture. High-quality dynamic microphones are expensive. The number of turns that can be placed on the moving coil is limited by space and the desire to keep the mass of the moving parts small. For this reason the resistance or impedance of the coil is limited in value to about 25 ohms. Microphones of higher impedance are manufactured, but these have step-up transformers normally built into the microphone case. Values of 600 ohms and 50,000 ohms are common, although other values are sometimes used. Since a transformer is used the d.c. resistance will be much less than the impedance. The use of a transformer also increases the voltage output or sensitivity. The movement of the diaphragm depends on the sound pressure, and hence this type of microphone is known as a pressure-operated type. For this reason the response of the microphone is approximately the same in all directions, what is called omnidirectional. This type of microphone is very

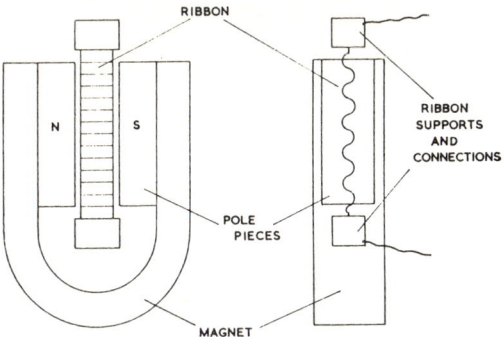

RIBBON

RIBBON
SUPPORTS
AND
CONNECTIONS

N   S

POLE
PIECES

MAGNET

FIG. 8.2. CONSTRUCTION OF RIBBON MICROPHONE

commonly used in domestic equipment and by professionals.

The principle of the ribbon microphone is shown in figure 8.2. A powerful magnet M has two pole-pieces N and S, and between them is the ribbon. The ribbon consists of a single strip of metal which is corrugated to make it more flexible. It is held in two insulated supports. If there is a DIFFERENCE in pressure between the two sides of the ribbon it will move and the movement will result in an e.m.f. being generated in the ribbon, due to the magnetic field. Because the ribbon will only move due to a pressure difference it is commonly known as a pressure gradient or velocity type of microphone. For this reason it does not have a uniform response, but has a maximum response when the sound waves are travelling in a line at right angles to the face of the ribbon. The response curve is, in fact, a figure 8 as shown in figure 8.3. This indicates that the response is a maximum along the line AD, but the response is zero for sound arriving

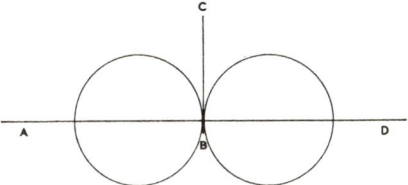

C

A   B   D

FIG. 8.3. POLAR DIAGRAM OF RIBBON MICROPHONE

along a line CB (in either direction). Obviously, the impedance of the ribbon is very small and so is the induced e.m.f., and hence a step-up transformer is always used. The results from such a microphone can be very good, but it is easily damaged. If blown into or used outside where the wind can blow on the ribbon, this may be displaced permanently, and the microphone is damaged. Accordingly it cannot be used outside.

The crystal microphone makes use of the piezoelectric property of certain materials, and has been described in connection with ceramic filters in Chapter 2. These materials have the property that, if a force is applied to them, an e.m.f. will be generated between, say, two conducting plates attached to them. One form of construction is given in figure 8.4, where a thin diaphragm D is used, connected to the piezoelectric crystal C, which operates by being bent. The microphone is operated by pressure of the sound waves, and hence has an approximately uniform response in all directions, *i.e.* it is omnidirectional. Its impedance is very high and must therefore feed an amplifier with a high input impedance, say 1 MΩ. It therefore is not suitable for use with the normal transistor amplifier. It has a reasonably uniform frequency response and is cheap.

The electrostatic or capacitor microphone is similar in basic construction to the electrostatic loudspeaker, and the principle is shown in figure 8.5. There is a fixed

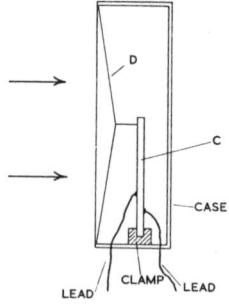

FIG. 8.4. ONE FORM OF CONSTRUCTION OF CRYSTAL MICROPHONE.

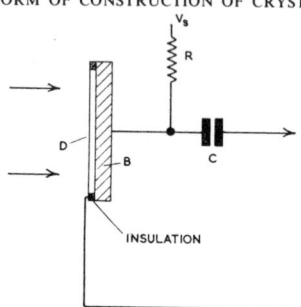

FIG. 8.5. PRINCIPLE OF ELECTROSTATIC MICROPHONE

conducting back plate B, and held close to it, but insulated from it, is a thin conducting diaphragm D. The back plate and diaphragm form the two plates of a capacitor, and when sound waves impinge on the diaphragm it vibrates and causes variations in capacitance. It does not generate an e.m.f. like the dynamic, ribbon and crystal microphones. However, if it is fed through a high resistor R, from a d.c. supply of, say, 100 V, then the changes in capacitance will cause changes in voltage across it, which can be fed to an amplifier through the d.c. blocking capacitor C. Due to the high impedance of the microphone it MUST feed an amplifier with a VERY high input impedance, the amplifier normally being built into the body of the microphone. The amplifier makes use of an F.E.T. (field effect transistor) which has a high input impedance. A disadvantage of this microphone is the need for a high voltage d.c. supply. When suitably manufactured such a microphone has an excellent frequency response and is often used as a standard. The normal microphone is pressure operated, and hence it is omnidirectional.

The need for a d.c. supply has been overcome by the use of what is called an electret microphone, the basic idea being shown in figure 8.6. In place of the conducting diaphragm an insulating diaphragm is used made of a plastic which has been permanently polarized. This means that it has a permanent electrostatic charge, corresponding to a permanent magnet. This will induce a charge in the base plate B, and when the diaphragm vibrates this will vary and an output voltage is produced. This is obviously much more convenient and is often used in domestic equipment. The frequency response of this type of microphone can be excellent. Again, it must feed a very high impedance amplifier, which may be built into the microphone case and might use an integrated circuit with an F.E.T. to get the high input impedance. This can be powered by a small battery of, say, $1\frac{1}{2}$ volts. The output impedance quoted for this type of microphone is that of the amplifier and can be any of the common values, e.g. 600 ohms.

Extremely small microphones are made using this principle, e.g. 10 mm diameter by 18 mm length, and are commonly being used in studios in the form of Lavalier

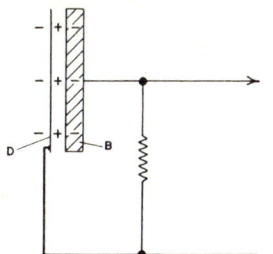

FIG. 8.6. PRINCIPLE OF ELECTRET MICROPHONE

microphones, which can be clipped to a speaker's clothes and be inconspicuous.

For many purposes an omnidirectional microphone is required so that sounds are picked up from all directions. However, a directional microphone is also often required so that unwanted sounds can be reduced. The common type of directional microphone has a cardioid shape of response, as shown in figure 8.7. The response from the back is small, and hence can be used to reduce interference sounds from the back.

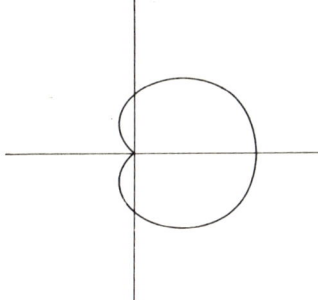

FIG. 8.7 POLAR DIAGRAM OF CARDIOID MICROPHONE

This characteristic is also used to prevent howling when microphones are placed near to loudspeakers. The microphone still has some response at the sides but this is less than that at the front. It is sometimes said that this response is obtained by combining the outputs of a ribbon and omnidirectional microphone, but in modern microphones it is obtained by using a suitable opening at the back of the diaphragm, of either a dynamic or an electrostatic microphone. The method of operation is involved and careful design is required to get good results. The cardioid shape shown in figure 8.7 is rather ideal, and this type of response is normally only obtained at middle frequencies. Its directional properties fall off at low frequencies. Supercardioid and ultra or hypercardioid microphones are made, which have a generally similar directional response. The ultra or hypercardioid has less response at the sides, but has a greater response from the back direction than a cardioid type.

For some purposes a much more directional microphone is required. A highly directional microphone can be produced by using a parabolic reflector behind a microphone, as in figure 8.8. The basic idea is that the reflector reflects the sound waves falling on it and concentrates them on the microphone. This results in increased sensitivity of the microphone (say by a factor of 20 dB) as well as making it highly directional. Unfortunately, the operation is quite involved and the frequency response depends on the size of reflector, but with any reasonable reflector size (say 60 cm diameter) the microphone sensitivity and directional properties drop off badly at low frequencies. The reflector may be made of metal or fibreglass. It is commonly used by bird recordists.

The reflector is inconvenient to use because of its size, and has been largely replaced, in the professional field, by the gun or rifle microphone. This consists of a tube

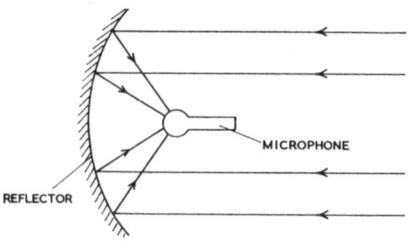

FIG. 8.8 USE OF REFLECTOR WITH MICROPHONE

(approximately 60–70 cm long and 2 cm diameter) with a number of slits in it connected to a dynamic or electrostatic microphone. The principle of operation is complex. Although this produces a directional microphone there is no increase in gain as with the reflector. The directional properties vary with frequency and, like the reflector, the directional property is not as good at low frequencies. It is normally used outside with a wind-shield, which is a long cylinder with a metal gauze outside and plastic foam lining. The gun microphone is more convenient than the reflector, and in general has a better frequency response, but is expensive.

Microphones are available with varying impedence values, the values being of some importance. Maximum power transfer will take place when the input impedance of the amplifier is equal to the output impedance of the microphone, but microphones are commonly used with amplifiers having an input impedance of 2 to 3 times their output impedance. This results in a higher voltage gain. (Some manufacturers quote the optimum load for their microphones rather than the output impedance). The available impedances cover a large range, but they can be classified into (a) low; (b) medium; and (c) high impedance. The low impedance microphones cover the range of, say, 15 to 100 ohms, the common values being 30 and 60 ohms. Medium impedance vary from, say, 100 to 1000 ohms, values of 200, 300 and 600 being common. High impedance cover, say, 2 kΩ to 100 kΩ, 50 kΩ being a common value. Crystal microphones and electrostatic microphones have not been included in this list. Crystal microphones have very high impedances and not generally used with transistor amplifiers. The electrostatic microphones have their own preamplifiers and the output impedance of these amplifiers can be made almost any value required. If a microphone is to be used a long way from the amplifier then a low impedance microphone should be used, because the capacitance between the leads of the cable will have less effect on the frequency response. There will also be less chance of interference being picked up. The microphone may be connected unbalanced as shown in figure 8.9 (a), in which case one side is at earthy potential. This is probably the most common arrangement with domestic equipment. The alternative is the balanced connection shown at (b).

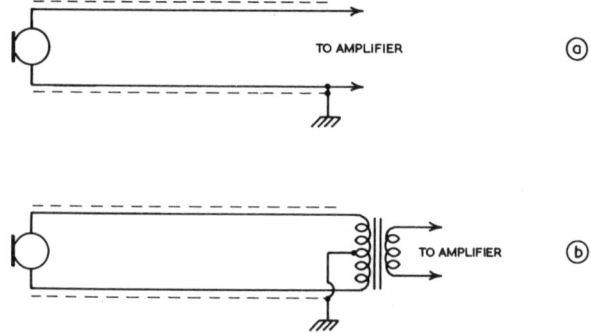

FIG. 8.9. CONNECTION OF MICROPHONE TO AMPLIFIER. (a) UNBALANCED; and (b) BALANCED

The leads are now balanced with respect to earth, and less interference is likely to be picked up, particularly when long leads are used. In all cases screened cable should be used (specially screened microphone cable is available) so that electrostatic pick up is reduced to a minimum. With medium, and especially high impedance microphones, long leads should not be used, and a screened lead is most important. Some microphones are fitted with a switch or links so that the output impedance can be changed (by using a transformer). If a microphone is to be used with an amplifier of unsuitable impedance then a matching transformer can be used.

The sensitivity of a microphone is generally important and there are a number of ways of expressing this which can make comparison difficult. It is sometimes quoted as the output in millivolts (usually at 1 kHz) for a sound pressure of 1 microbar, which for all practical purposes, is the same as a pressure of 1 $dyne/cm^2$. Thus the sensitivity might be given as, say 0.2 mV/μ Bar. The sensitivity is usually quoted as the open-circuit voltage, and will be less when feeding the input impedance of an amplifier. The sensitivity may also be expressed in terms of a pressure measured in Newton/square metre. Since 1 microbar $= 0.1$ Newton/square metre, a sensitivity of 0.2 mV/μ Bar becomes 2 mV/Newton/square metre (2 mV/N/m^2). The sensitivity is also quoted in dB, but this must be relative to some standard. It is unfortunate that different standards are used and are not always stated. One standard is in terms of power, the 0 dB figure corresponding to 1 mW/10μ bar or 1 mW/1μ bar, the former corresponding to 1 mW/Newton/m^2. Other standards are 1 volt/μ bar or 1 volt/10μ bar, the latter corresponding to 1 volt/Newton/m^2. The quoted figure when the standard is 1 volt/10μ bar will be $+20$ dB compared with that when the standard (corresponding to 0 dB) is 1 volt/μ bar. Because of all these standards care is necessary when comparing the sensitivity of microphones. The sensitivity in terms of open-circuit voltage will be greater the higher the impedance of the microphone. For example, it might be -87 dB (standard 1 volt/μ Bar) at 25Ω, -75 dB at 200 Ω and -52 dB at 50,000 ohms, so again care is necessary in comparing microphones. The greater sensitivity of a high-impedance microphone can, of course, only be made use of if a suitable amplifier with a higher impedance is available. The sensitivity of microphones of the same impedance does not vary greatly; in general the better the quality of the microphone the less the sensitivity, although this may not always apply.

CHAPTER 9
STEREOPHONIC SOUND TRANSMISSION

IN monophonic reproduction one microphone (more than one may be used but with the outputs combined) is used at the studio and a single speaker is used for reproduction. Thus we are reproducing the sound pressure at any instant. The result we get is not the same as if a person were in the studio because he has two ears. This enables him to tell in which direction the sound is coming but, of course, this information is lost when we reproduce a signal from a single loudspeaker. The manner in which the ears are able to detect direction is complex and not fully understood. If the sound source is to one side one ear will hear the sound of greater intensity than the other, and also one ear will hear the sound a fraction of a second earlier than the other. In some way the brain is able to interpret this information and say from which direction the sound has come. The idea of a stereophonic system is to supply this additional information by using two channels, which correspond approximately with the signals received by the two ears. The basic principle is shown in figure 9.1. At the studio two microphones are used which are spaced apart so that two different

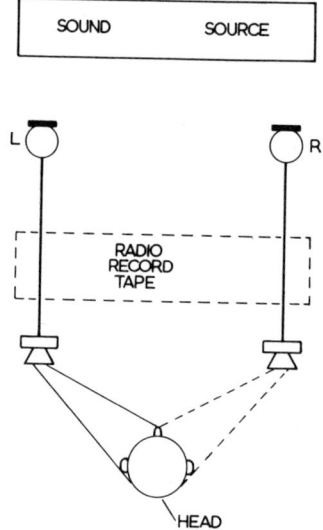

FIG. 9.1.   PRINCIPLE OF STEREO REPRODUCTION

signals are picked up, these being referred to as left-hand (L) and right-hand (R) signals. The two signals are then fed to two loudspeakers, which we will assume are spaced apart by the same distance as the microphones. The signals from the microphones can be fed by radio transmission or by means of a record or by tape. The listener then sits in a position as shown and then receives sounds from both loudspeakers. In practice this system does not work well. It is possible to tell whether sound source is to the left or right, but little sound appears to come from the middle and this is known as "hole in the middle". The defect can be overcome by using two directional microphones (such as cardioid) placed in a central position, but at right angles to each other, as shown in figure 9.2. Thus the sound source on the left is picked up mainly by the left-hand microphone and fed to the left-hand speaker. The listener then hears the sound from the left-hand speaker, and so on. The exact analysis is complex and will not be considered here. For best results the two channels should be similar

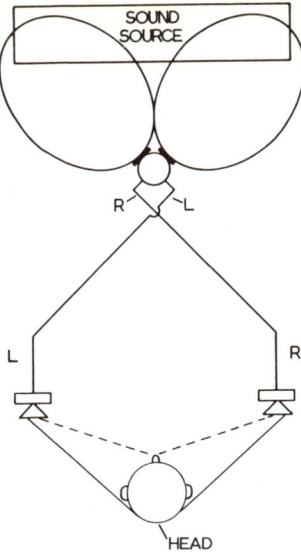

FIG. 9.2. STEREO REPRODUCTION USING CROSSED MICROPHONES

and similar speakers should be used. The loudspeakers should be spaced apart by 6 to 10 feet, and the listener should sit an equal distance from each, the angle formed by the speakers being about 60° to each loudspeaker from the centre line or 120° between loudspeakers. Obviously it is not always possible to fulfil all these conditions, particularly if a number of people are listening, but the more one is off centre the less the true stereo effect. It might be thought that with a solo player there would be no point in stereo. This is not so because not all the sound comes direct; much comes by reflections from walls, etc., and it is these reflections that give realism to the reproduction.

The stereo system does not give any information concerned with sound from the rear of the listener. This can be overcome by using four channels, quadraphony as it is called. This will be considered in Chapter 12.

We are now mainly concerned with the way in which a stereo signal is transmitted by radio. More and more radio transmissions on v.h.f. are in stereo.

It will be assumed that at the studio we have two audio signals available which we will call left (L) and right (R) signals. It is required to transmit these two signals so that they can be reproduced on two speakers. The simplest way might be to transmit one signal on one channel and the other on another channel, but this has serious disadvantages.

(1) It would be difficult to maintain a correct balance between the two signals, and any variation in the magnitude of one signal relative to the other causes a movement of the stereo image.

(2) It would occupy twice the bandwidth of mono, or, put another way, we could only have half the number of programmes for a given band of frequencies.

(3) In order to receive a mono signal it would be necessary to receive both channels and add them. In other words, the system would not be compatible with the mono transmissions.

It is therefore necessary to develop some system which can be transmitted within the same bandwidth as mono and is compatible, *i.e.* it can be received by an unmodified mono receiver and give a satisfactory mono reproduction. The system should also be such that variations in signal strength do not upset the RELATIVE amplitudes of the two signals. Some coding system must be used and that used in this country (and in other countries) is called the Zenith-GE multiplex system or pilot tone system.

In this system instead of transmitting the L and R signal directly, they are first coded to give:

(a) the sum signal L + R; and (b) the difference signal L − R.

The L + R signal is the same as the mono signal and is transmitted in the normal way. This signal will operate a mono receiver and the system is therefore compatible. The L − R signal must be transmitted in such a way (a) that it does not interfere with the mono receiver; and (b) that it can be recovered in a stereo receiver. The L − R signal is used to modulate a subcarrier outside the audio band so that it does not cause interference on a mono receiver.

The frequency of the subcarrier is 38 kHz and this is amplitude modulated with the L − R signal. However, normal amplitude modulation is not used, but what is called "suppressed carrier modulation". This will first be explained. Figure 9.3 shows normal amplitude modulation of a carrier. During the time A to B there is

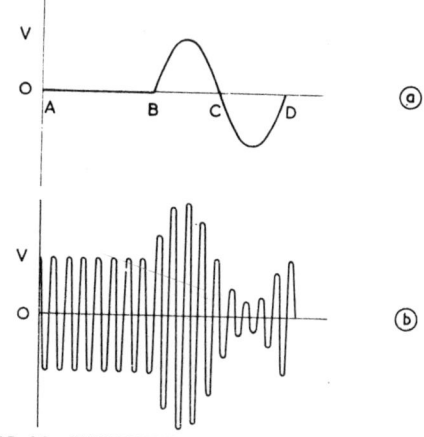

FIG. 9.3.   WAVEFORMS OF AMPLITUDE MODULATION

no modulation, and hence the carrier is of constant amplitude. From B to C the modulating voltage is positive and hence the carrier amplitude is increased in proportion to the modulating voltage. From C to D the modulating voltage is negative and hence the amplitude of the carrier is reduced. It can, of course, be shown that this is equivalent to transmitting a carrier and two sidebands, as is shown in figure 9.4. The sidebands are spaced from the carrier by the modulating frequency, and at 100% modulation each sideband has an amplitude half that of the carrier. The idea of

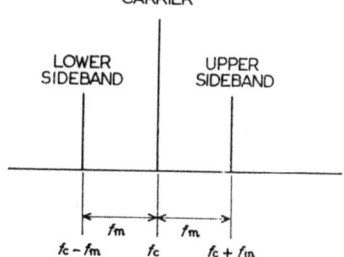

FIG. 9.4.   SIDEBANDS PRODUCED AS A RESULT OF AMPLITUDE MODULATION

suppressed carrier modulation is to remove the carrier so that only the two sidebands remain. The magnitude of the sidebands is proportional to the magnitude of the modulating voltage and hence is zero with no modulation. This is quite different to normal amplitude modulation, where there is a constant amplitude carrier independent of the percentage modulation. This difference is of vital importance as we shall see later. The waveform of suppressed carrier modulation is different, and is shown in figure 9.5. From A to B the modulating voltage is zero, and so there is nothing from the modulator. From B to C the modulating signal is positive, and

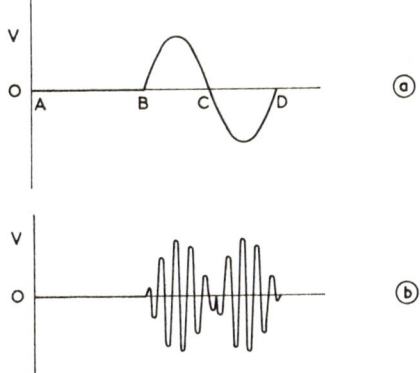

FIG. 9.5. WAVEFORMS OF SUPPRESSED CARRIER MODULATION

there is an output from the modulator (at carrier frequency) proportional to the amplitude of the modulating voltage. From C to D the modulating voltage is negative, and the modulator output again produces a signal proportional to the amplitude of the modulating voltage. It should be noted that the outline of the modulated wave is no longer a replica of the modulating voltage, as it was in figure 9.3. It can be shown that when the modulating voltage changes from positive to negative, the phase of the carrier reverses, as shown in figure 9.5. It is this phase reversal that enables the demodulator to detect whether it is a positive or a negative half-cycle.

We must now look at the stereo frequency spectrum produced which is shown, for the general case, in figure 9.6. The L + R signal is, of course, like a mono signal, and we will assume that this extends from the lower limit of 20 Hz to an upper limit of 15 kHz. The missing subcarrier would be at 38 kHz, and therefore the sidebands extend in both directions from this frequency. Assuming the L − R signal to have the same bandwidth, the upper sidebands will extend from 38,000 + 20 = 38,020 Hz to 38,000 + 15,000 = 53,000 Hz or 53 kHz. The lower sidebands will go from 38,000 − 20 = 27,980 to 38,000 − 15,000 = 23,000 Hz or 23 kHz. This is shown in the figure. In

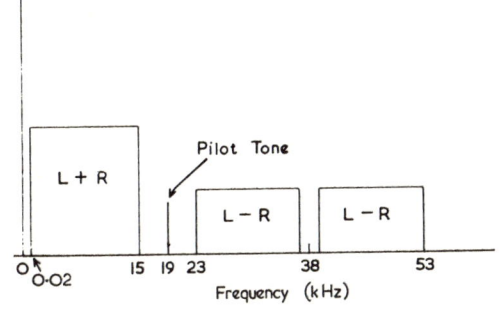

FIG. 9.6. FREQUENCY SPECTRUM OF STEREO SIGNAL (GENERAL)

order that the suppressed carrier modulation can be recovered, it is necessary to reinsert the subcarrier at the receiver or have available a subcarrier frequency equal and in phase with that at the transmitter. To do this a pilot tone is transmitted at half the subcarrier frequency, *i.e.* 19 kHz. At the receiver this is doubled in frequency to obtain the subcarrier frequency. The magnitude of the pilot tone is kept small, 9%, to prevent using excessive bandwidth (see later). One might say: Why not transmit 9% of the subcarrier, as this would sound simpler. If this were done it would be very difficult to pick out this subcarrier because there are sidebands only 20 Hz away on each side and an extremely selective filter would be required. If figure 9.6 is examined it will be seen that there is a vacant space of 4 kHz each side of the pilot tone, and hence there is no difficulty in picking out this frequency. In order that the subcarrier at the receiver is of the correct phase, there must be a definite relationship between the pilot tone and subcarrier, and this is given in figure 9.7. The phase relationship is such that, when the pilot tone is zero, the subcarrier is zero and going in a positive direction.

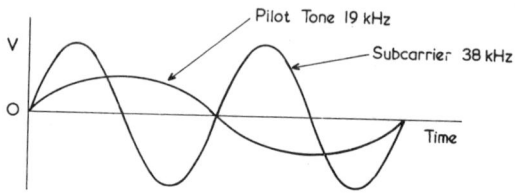

FIG. 9.7. PHASE RELATIONSHIP BETWEEN SUBCARRIER AND PILOT TONE

It is important to say something about the nature of the L and R signals since we are adding and subtracting them, and they are alternating quantities. Both L and R are normally complex signals, say from two microphones in front of an orchestra. The waveforms of the signals will be complex and continually changing. The important point is that there is no relationship between the two signals, either as regards frequency or phase. For example, one microphone may be receiving a signal from an instrument on the left-hand side, while the other microphone is receiving a sound signal from a quite different instrument on the right-hand side. Since there is no relationship between them, at some instant the peak amplitude of one will occur at the same instant as the peak amplitude of the other, and the resultant will be the sum of the peak amplitudes. Thus, when we talk about adding we are adding instantaneous values. Thus, if the peak value of the L signal is numerically equal to x and the peak value of the R signal is y then the L + R signal will be the sum of the peak values, *i.e.* $x + y$. The L and R signals are those at the transmitter after pre-emphasis. L and R are limited in value to ± 1 and the sum signal is made $(L + R)/2$ so that if L and R are $+1$ then $(L + R)/2$ is $+1$, and if L and R are -1 then $(L + R)/2$ is -1. Thus the limits of the $(L + R)/2$ signal is ± 1. In a similar way the difference signal is made $(L - R)/2$ which also has maximum values of ± 1. It is useful to see how the values of $(L + R)/2$ and $(L - R)/2$ signals vary as L and R vary. This is shown in Table 9.1, where L has been maintained constant at $+1$ while R is varied from $+1$ to -1. An important factor appears from this table, that is, when $(L + R)/2$ is large $(L - R)/2$ is small, and *vice versa*. In fact, it will be seen that the sum is a constant. Several extreme cases will now be considered.

(1) L = R = ± 1

This corresponds to a mono signal. The sum signal $(L + R)/2 = \pm 1$ and this is used to modulate the normal v.h.f. carrier. However, instead of using 100% modulation the percentage modulation is reduced to 90% so as to leave room for the pilot tone. 100% modulation results in a deviation of ± 75 kHz and with 90% modulation the deviation is $0.9 \times 75 = 67.5$ kHz. The pilot tone modulates the carrier by only 9%.

TABLE 9.1

L	R	(L+R)/2	(L−R)/2
1	1	1	0
1	0·8	0·9	0·1
1	0·6	0·8	0·2
1	0·4	0·7	0·3
1	0·2	0·6	0·4
1	0·0	0·5	0·5
1	−0·2	0·4	0·6
1	−0·4	0·3	0·7
1	−0·6	0·2	0·8
1	−0·8	0·1	0·9
1	−1	0·0	1·0

The (L − R)/2 signal is zero, and hence there are no sidebands of the subcarrier. Thus we have:

Total percentage modulation = 90% + 9% = 99%

$$\underset{\substack{\dfrac{L+R}{2}}}{|} \qquad \underset{\substack{\text{pilot}\\ \text{tone}}}{|}$$

The spectrum for this condition is shown in figure 9.8.

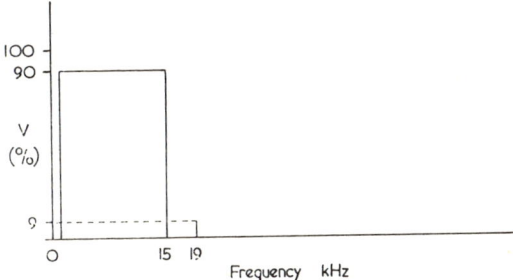

FIG. 9.8.   FREQUENCY SPECTRUM WHEN L = 1 AND R = 1

Figure 9.9 shows at (a) the resulting waveform without the pilot tone, and at (b) with the pilot tone. The top waveform is the output from the coder and the lower waveform is the audio frequency input (in these diagrams 1000 Hz). Since there is no frequency relationship between the pilot tone and the modulation the pilot tone runs round the waveform and is not stationary.

(2) L = +1   R = −1

In this case (L + R)/2 = 0 and (L − R)/2 = 1.

There is, therefore, no modulation due to the (L + R)/2 signal. The (L − R)/2 will produce sidebands of 50% because the modulation is 100%. The actual modulation

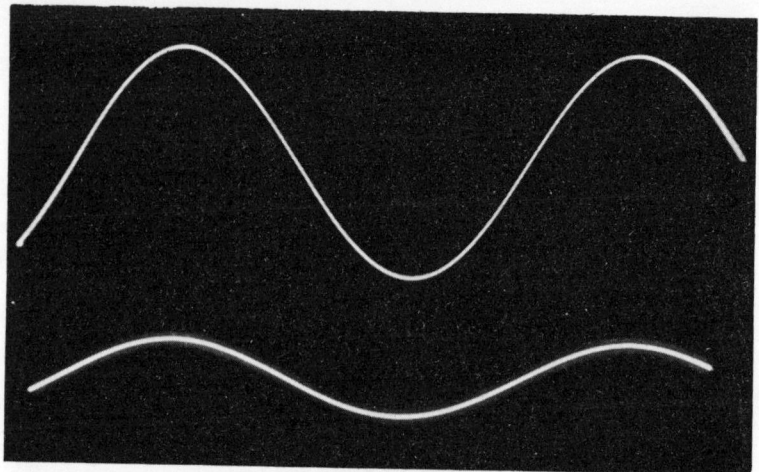

(a) WITHOUT PILOT TONE

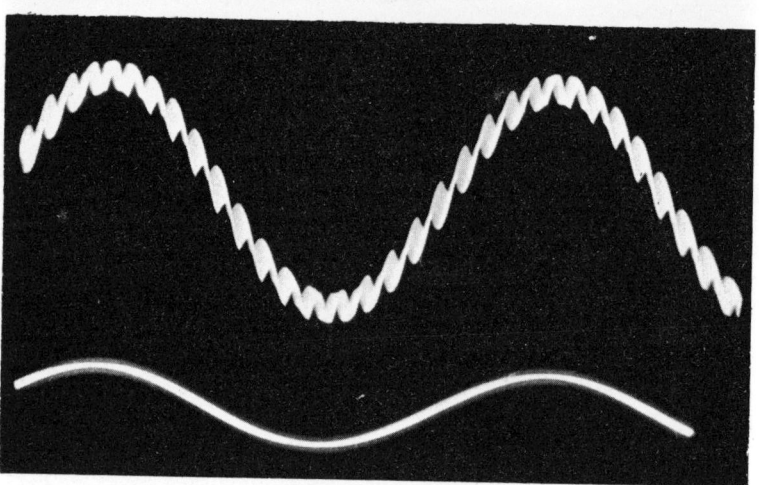

(b) WITH PILOT TONE

FIG. 9.9. WAVEFORMS OF STEREO SIGNALS WHEN L = 1 AND R = 1. UPPER WAVEFORM IS OUTPUT AND LOWER THE INPUT.

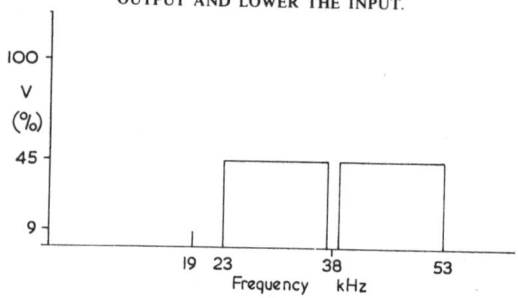

FIG. 9.10. FREQUENCY SPECTRUM WHEN L = 1 AND R = −1

of the v.h.f. carrier will be $0.9 \times 50 = 45\%$, as seen in figure 9.10. The total modulation is therefore:

$$0 + 45\% + 45\% + 9\% = 99\%$$

$$\underset{\substack{\text{lower}\\\text{sideband}}}{\underbrace{\frac{L-R}{2}}} \quad \underset{\substack{\text{upper}\\\text{sideband}}}{\underbrace{\frac{L-R}{2}}} \quad \underset{\substack{\text{tone}}}{\text{pilot}}$$

It is seen that the maximum percentage modulation is the same—almost 100%.

The corresponding waveform diagram is shown in figure 9.11. At (a) without pilot tone and at (b) with pilot tone. This waveform is that expected by suppressed carrier modulation of the subcarrier. If $L = -1$ and $R = +1$ similar results are, of course, obtained.

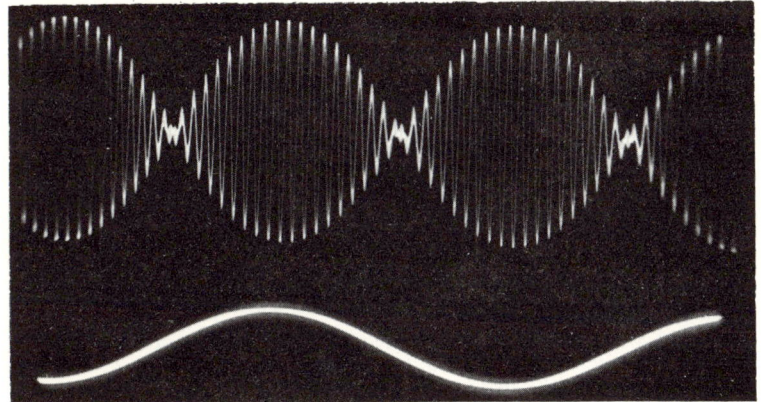

(a) WITHOUT PILOT TONE

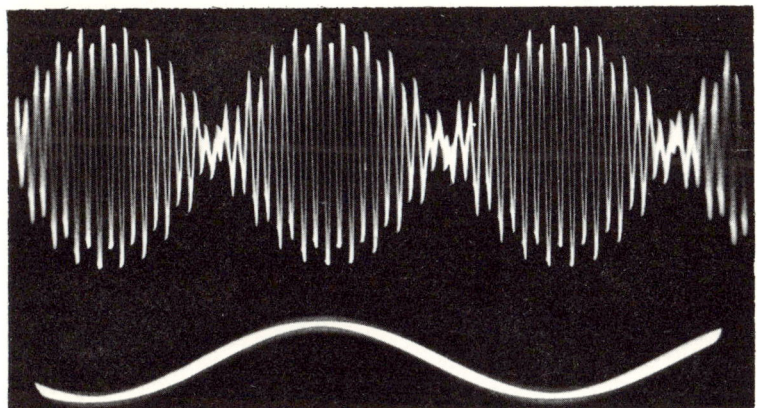

(b) WITH PILOT TONE

FIG. 9.11. WAVEFORMS OF STEREO SIGNALS WHEN L = 1 AND R = −1. UPPER WAVEFORM IS OUTPUT AND LOWER THE INPUT.

(3) L = 1   R = 0

This corresponds to a signal in only one microphone, the left one.

In this case $(L + R)/2 = 0.5$ and $(L - R)/2 = 0.5$.

The $(L + R)/2$ signal will modulate the v.h.f. carrier not 50% but $0.9 \times 50 = 45\%$ as shown in figure 9.12. If 100% modulation of the subcarrier takes place then each sideband will be half the amplitude of the carrier, but in this case there is only 50%

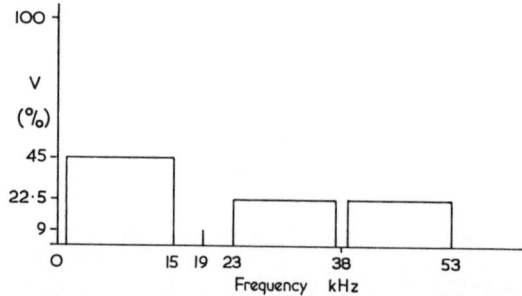

FIG. 9.12. FREQUENCY SPECTRUM WHEN L = 1 and R = 0

modulation, and therefore each sideband would have an amplitude of 25%. Allowing for the fact that only 90% modulation is used, each sideband becomes $0.9 \times 25 = 22.5\%$, as shown in the figure. The total percentage modulation is therefore:

$$45\% \quad + \quad 22.5\% \quad + \quad 22.5\% \quad + \quad 9\% \quad = \quad 99\%$$

$\dfrac{L + R}{2}$	$\dfrac{L - R}{2}$	$\dfrac{L - R}{2}$	pilot tone
	lower sideband	upper sideband	

It is seen that the maximum percentage modulation is again approximately 100%.

The waveforms corresponding to this condition are shown in figure 9.13. This waveform needs some explanation. It is the resultant of the audio frequency and the modulated subcarrier, in other words it is the sum of the waveform of figure 9.9 and figure 9.11. Since under these conditions the two sets of waveforms are equal, the waveform is either all positive or all negative during each half-cycle of modulation.

If L = 0 and R = 1 a similar result is obtained.

It is important to stress that these are maximum values, and that extreme cases have been taken. In practice, for most of the time L and R will be MUCH less than unity, and hence the deviation will be much less than 75 kHz.

BANDWIDTH

This type of stereo signal cannot be used to amplitude modulate a medium wave transmitter because of the high modulating frequencies. The 53 kHz signal would produce upper and lower sidebands 53 kHz from the carrier and require a bandwidth of $2 \times 53 = 106$ kHz. Such a bandwidth is just not available, consequently this stereo system can only be used on v.h.f.

If we consider a mono transmission, it was shown in Chapter 1 that with a maximum modulating frequency of 15 kHz and a deviation of 75 kHz the modulation index is $75/15 = 5$. This produces 16 sidebands and gives a total bandwidth of $16 \times 15 = 240$ kHz. Again it must be remembered that this will only occur when there is 100% modulation at a frequency of 15 kHz, which is most unlikely to occur. The bandwidth for other frequencies is generally less.

In a stereo transmission where $(L - R)/2 = 0$ (really a mono signal) the deviation is 67.5 kHz [see case (1)]. The modulation index is now $67.5/15 = 4.5$. This gives 14

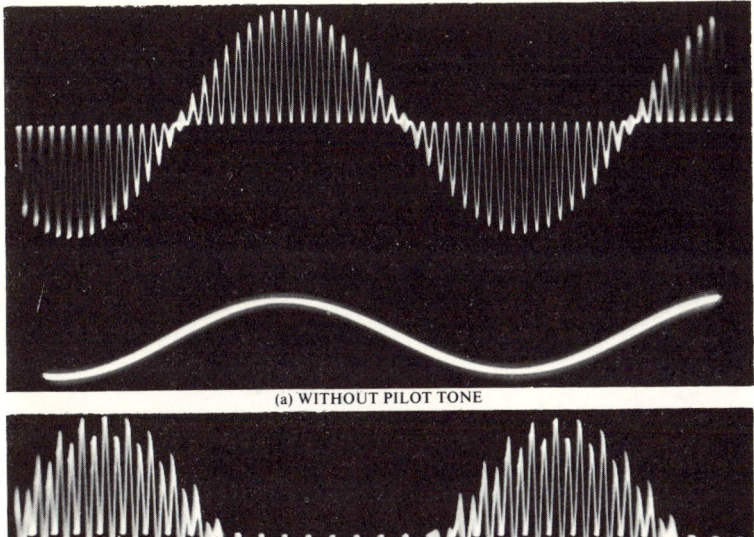

(a) WITHOUT PILOT TONE

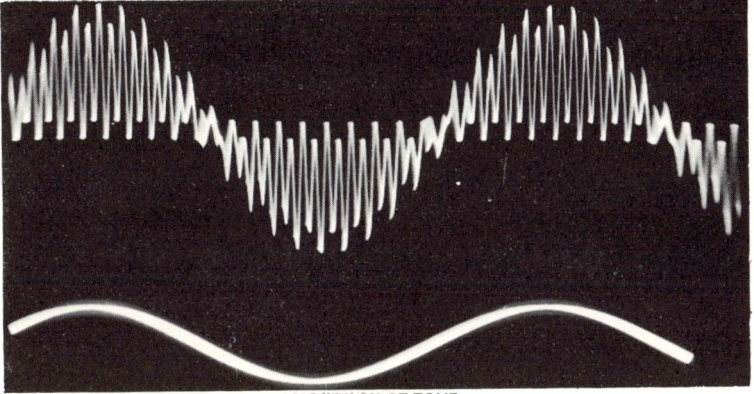

(b) WITH PILOT TONE

FIG. 9.13. WAVEFORMS OF STEREO SIGNALS WHEN L = 1 AND R = 0. UPPER WAVEFORM IS OUTPUT AND LOWER INPUT.

significant sidebands, and a bandwidth of $14 \times 15 = 210$ kHz, slightly less than the normal mono transmission. The effect of the pilot tone has been neglected.

If we now consider the case where $(L + R)/2 = 0$ or $(L - R)/2 = 1$ this produces the two sidebands of the subcarrier of 45% and these extend from 23–53 kHz (apart from the small gap in the centre). Consider the maximum frequency of 53 kHz. The deviation will be $45 \times 75 = 34$ kHz and the modulation index becomes $34/53 = 0.64$. The number of significant sidebands is now 4, and the bandwidth is $4 \times 53 = 212$ kHz. It can be shown that this is not the worst case and that a larger bandwidth occurs for a frequency of 42 kHz when there are 6 sidebands. The bandwidth is then $6 \times 42 = 252$ kHz. This is in fact very little different from that for a mono transmission. However, it is better to design the receiver to have adequate bandwidth for a stereo signal say 250–300 kHz. These calculations are a little academic as they assume maximum modulation under special conditions, and maximum modulation is likely to occur only occasionally on very loud passages of music or speech. The elimination of sidebands of small amplitude is likely to have little effect.

DEMODULATOR

The demodulator at the receiver must now deal with modulation frequencies going up to 53 kHz, instead of only 15 kHz on mono, and it must be designed to deal with them. The de-emphasis must NOT be used before the decoder, or the higher frequencies corresponding to the $(L-R)/2$ signal will be greatly attenuated and will be too small to operate the decoder. De-emphasis must, of course, be applied to the L and R signals after they have been recovered in the decoder.

ALTERNATIVE APPROACH

In order to explain the operation of the most common type of decoder it is desirable to look at the system of transmission in a different way. Suppose we consider two microphones M_1 and M_2 as in figure 9.14, picking up L and R signals. These signals are fed through amplifiers A_1 and A_2 to switch S_1, which connects them

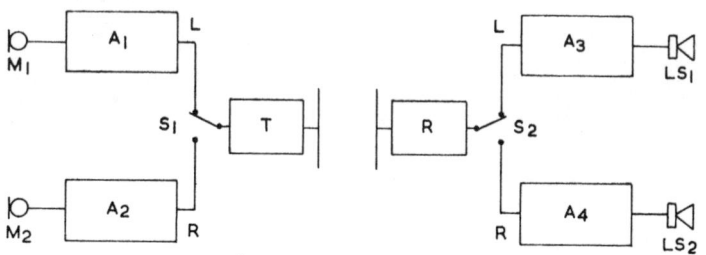

FIG. 9.14. IDEA OF SIGNAL SWITCHING

alternately to the transmitter T. The output of the receiver R is connected to a similar switch S_2 so that the signal is switched first to amplifier A_3 and loudspeaker LS_1 and then to A_4 and LS_2. The two switches are operated in synchronism at the subcarrier frequency of 38 kHz. Thus M_1 is connected to LS_1 for a brief instant and then M_2 is connected to LS_2 for a brief instant. Thus we have a switching system which is sometimes called a "time division multiplex system", because the two signals share the common link on a time basis.

Now consider some waveforms as shown in figure 9.15. At (a) is shown the L signal, which, for convenience, has been taken only as a sine wave. At (b) is shown the R signal, and again, purely for simplicity, this is taken as a sine wave of twice the frequency of the L signal. At (c) the sum signal $L+R$ is shown and (d) the difference signal $L-R$. Our switching system is therefore switching between the L and R signals at 38 kHz, and is shown at (e). From A to B the output is due to the L signal, while from B to C it is due to the R signal, and so on. If we pass the waveform of figure 9.15 (e) through a low-pass filter, so as to remove the 38 kHz carrier, we shall obtain the average of the signal which is shown dotted. Since at one instant the signal is L and at the other instant R, the average is $(L+R)/2$. It will be seen that the dotted curve is the same shape as the $L+R$ signal. Suppose that we pass the waveform through a high-pass filter, which will remove the low frequencies, corresponding to L and R. The result is at (f). This is the amplitude of the square wave 38 kHz signal. Since at one instant the signal is L and at the other instant R, the difference (which is the amplitude of the square wave) must be $L-R$. It is seen that (f) corresponds to the suppressed carrier modulation of $L-R$ signal. Thus, the $L-R$ signal can be recovered. In fact, the result of this switching produces the same waveform, i.e. the sum of $(L+R)/2$ and a subcarrier suppressed carrier modulated with $(L-R)/2$, the only difference being that in the switching case the carrier is of square waveform and not sinusoidal. The switching system is not used to produce the signal at the transmitter, but the switching idea is used in one type of decoder to be described later.

STEREO DECODERS

There are several types of decoder, but only four will be described: the matrix decoder, the switching decoder and two integrated circuit decoders. The decoder

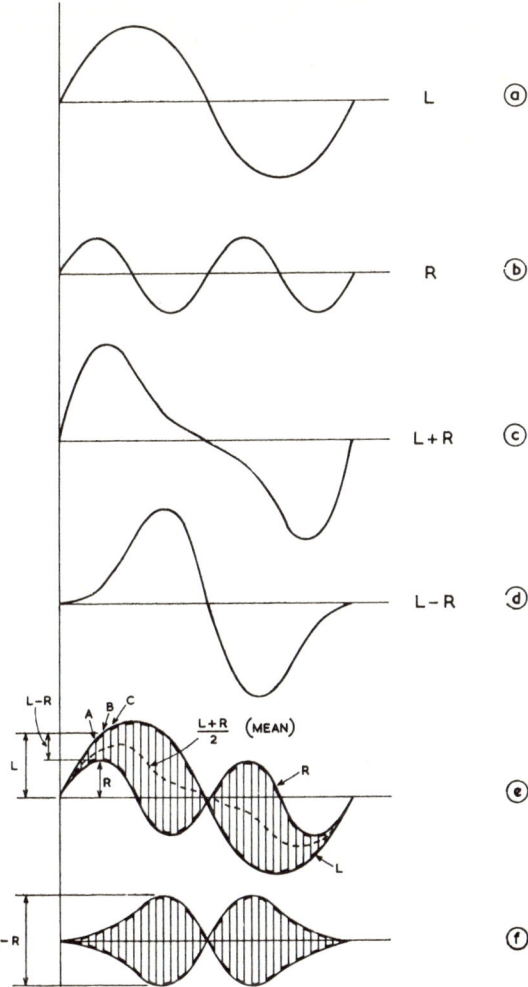

FIG. 9.15. STEREO WAVEFORMS USING SWITCHING IDEA

has three functions:

(a) Pick out the 19 kHz pilot tone and, from this generate the subcarrier at 38 kHz, which is required by the subcarrier demodulator.

(b) Demodulate the $(L - R)/2$ signal.

(c) Combine the $(L + R)/2$ and $(L - R)/2$ signals to obtain the L and R signals.

Functions (b) and (c) may be combined.

MATRIX DECODER

This decoder is described as it appears to be the obvious arrangement, but it is not very commonly used in practice as it has no advantage compared with the others.

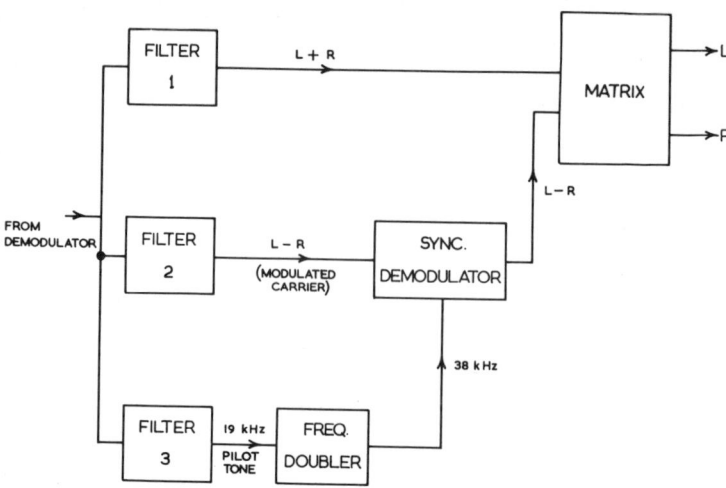

FIG. 9.16. BLOCK DIAGRAM OF MATRIX DECODER

A block diagram is shown in figure 9.16. The output from the demodulator is fed to the three filters. Filter one passes 20 Hz to 15 kHz and the output is therefore the $(L + R)/2$ which is fed into the matrix. Filter 2 picks out the range of frequencies 23 to 53 kHz, *i.e.* the sidebands of the $(L - R)/2$ signal, feeds them to the synchronous demodulator which is basically a multiplier. From this is obtained the $(L - R)/2$ signal which is fed into the matrix. The matrix combines the $(L + R)/2$ and $(L - R)/2$ and produces the L and R signals. By adding $(L + R)/2$ and $(L - R)/2$ we obtain $(L + R)/2 + (L - R)/2 = L$. By subtracting the two signals we get $(L + R)/2 - (L - R)/2 = R$. Filter 3 is tuned to 19 kHz and picks out the pilot tone, which is fed into the frequency doubler. The output of 38 kHz then feeds the subcarrier demodulator. Instead of using a multiplier for the subcarrier demodulator the arrangement of figure 9.17 may be used. In this case the subcarrier from the frequency-doubler is fed to the adder, where it is added

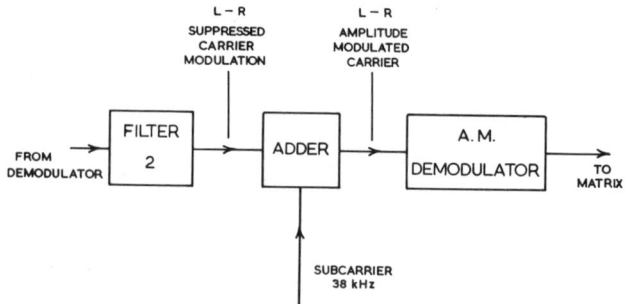

FIG. 9.17. ALTERNATIVE METHOD OF DEMODULATING SUBCARRIER

to the sidebands of the $(L - R)/2$ signal. In other words the carrier has been reinserted and the result is a normal amplitude modulated carrier. This can be demodulated in a normal amplitude demodulator (envelope detector).

SWITCHING DECODER

This is the most common type of decoder. Using the alternative approach to the stereo signal, what we need at the decoder (see figure 9.14) is a switch operating at

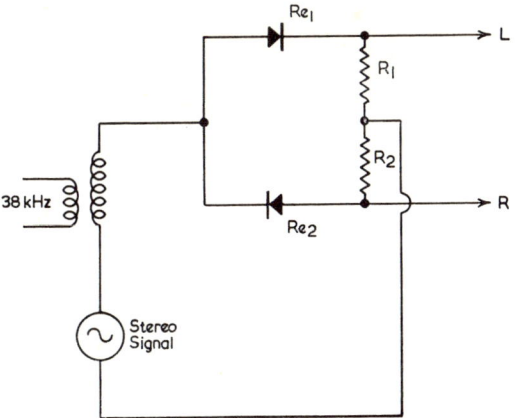

FIG. 9.18 BASIC STEREO SWITCHING CIRCUIT (UNBALANCED)

38 kHz, which will switch the incoming signal between the L and R outputs. This switch must obviously be electronic, and a simple circuit is given in figure 9.18. It will be assumed that a subcarrier frequency is available for operating the switch. Consider first the operation when the stereo signal is zero. On one half-cycle the 38 kHz voltage will cause a current to flow in Re_1 and R_1, and produce half-cycles of voltage across R_1. On the other half-cycle current will flow in Re_2 and R_2 and similar half-cycles will be produced across R_2. After smoothing there will be steady d.c. voltages produced across R_1 and R_2. Now suppose that the stereo signal is added in series with the switching signal, as shown, and that the stereo signal is smaller than the switching signal. On one half-cycle Re_1 is conducting and hence the stereo signal is fed to the L output, and will be superimposed on the voltage across R_1 due to the switching signal. Since the latter is of steady value it can be removed by a simple d.c. blocking capacitor. During this time Re_2 is reverse biased, and hence no signal is fed to the R output. On the next half-cycle Re_2 is conducting and, in a similar way, the stereo signal is fed to the R output. During this period Re_1 is reverse biased and therefore no signal is fed to the L output. Thus the stereo signal is fed alternately to the L and R outputs at 38 kHz as required.

The simple half-wave circuit has disadvantages:

(a) There will be large 38 kHz signals produced across R_1 and R_2 due to the half-cycles of the switching signal, and this large amplitude is difficult to remove.

(b) The smoothed voltage across R_1 and R_2 will only be of constant value if the magnitude of the 38 kHz signal is constant. If this is amplitude modulated for any reason then it will produce an output in both channels.

It is better to use a full-wave rectifier circuit or balanced circuit, and one is shown in figure 9.19. In this circuit $R_5 = R_6$ and $R_1 = R_2 = R_3 = R_4$. The resistors R_1 to R_4 are only added to balance the diodes Re_1 to Re_4 and are not essential. They will be small in value and small compared with R_5 and R_6. First, consider the operation with no stereo signal. During the first half-cycle of the 38 kHz switching waveform A will be positive with respect to B, and C will be negative with respect to B. A current will flow from A through R_1, Re_1, Re_2 and R_2 to point C. Since the circuit is symmetrical there will be no voltage produced across R_5. On the other half-cycle C is positive with respect to A and a current flows through R_4, Re_4, Re_3 and R_3. Again, since the circuit is symmetrical there will be no voltage across R_6. During this half-cycle Re_1 and Re_2 are reverse biased, and hence non-conducting. Thus on one half-cycle point B is connected to R_5 and on the other half-cycle B is connected to R_6. The stereo signal is added, as shown, and on one half-cycle the stereo signal is connected through

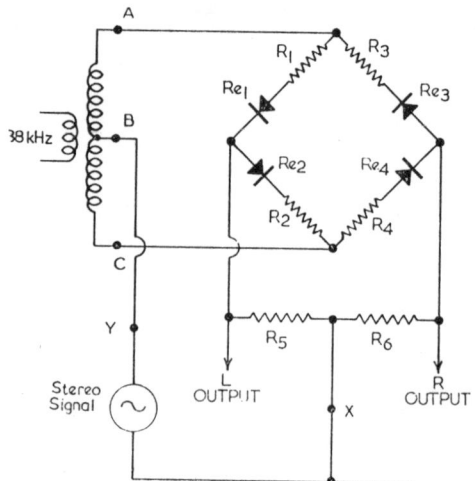

FIG. 9.19.   BALANCED SWITCHING CIRCUIT

Re_1 and Re_2 to the L output (across R_5) and on the other half-cycle through Re_3 and Re_4 to the R output (across R_6).

    Due to the balanced nature of the circuit there is no appreciable output at 38 kHz at the L and R outputs, which is a considerable advantage. Any variation in the magnitude of the switching signal (provided it is greater than the stereo signal) will not appear as a signal in the output and will have little effect on the magnitude of the L and R signals. Instead of applying the stereo signal to point Y it can be connected to point X, and point Y connected to the earthy line. The circuit drawn as in figure 9.19 may be drawn as in figure 9.20, which is electrically identical. Resistors R_1–R_4 may be omitted, but they help to balance out any differences in the characteristics of the diodes, and therefore make the circuit more symmetrical.

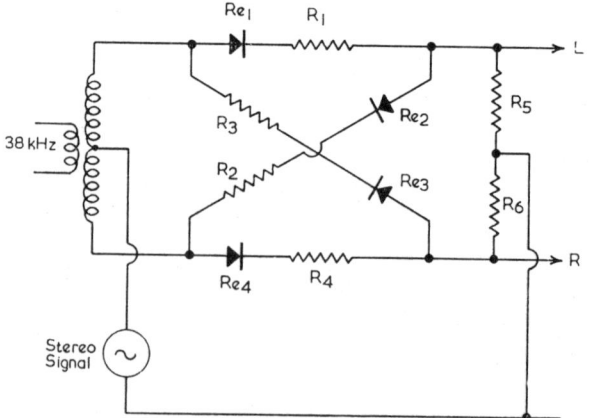

FIG. 9.20.   BALANCED SWITCHING CIRCUIT, ALTERNATIVE METHOD OF DRAWING

    Unfortunately, the switching decoder produces some crosstalk between the two outputs, *i.e.* the left-hand output contains some of the R signal, and the R output contains some of the L signal. This is due to the fact that the modulation at the coder is by a sinewave, whereas the switching is by a square wave. Consider figure 9.21 which shows the L signal at (a). Assume that the L signal is of magnitude 1 and the R

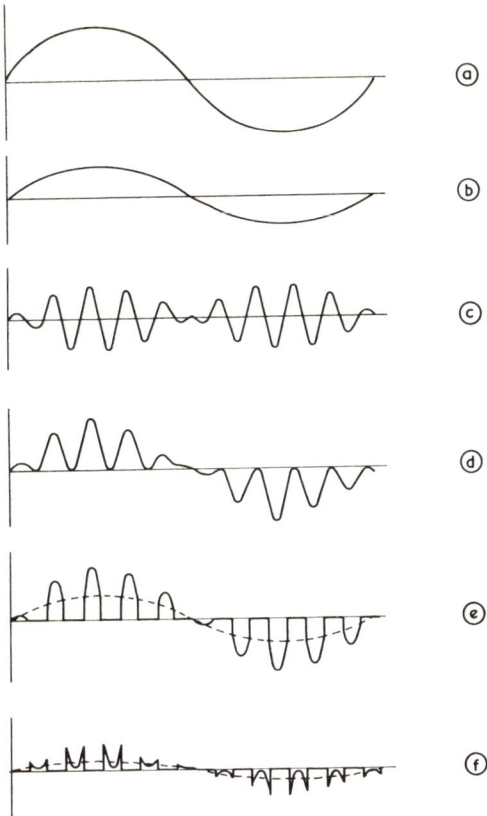

FIG. 9.21. PRODUCTION OF CROSSTALK IN SWITCHING DECODER

signal is zero. Thus, the $(L + R)/2$ signal becomes $L/2$ and is shown at (b). The difference signal $(L - R)/2 = L/2$ (since $R = 0$), and it is this signal which modulates the sub-carrier and will produce an output as at (c). The overall signal will be as at (d). The action of the switching decoder is shown at (e) and (f). During one half-cycle the signal at (d) is switched to the L output with the result shown at (e). During the other half-cycle it is fed to the R output as at (f). Ideally there should be no R output because the R signal was zero. When the carrier component of (e) is smoothed out, as shown dotted, it is seen that the output corresponds to the L input as required. Again, when the signal shown at (f) is smoothed there will be an output which is shown dotted. This corresponds also with the L input. Thus there is an L signal in the R output, being known as crosstalk. It is seen that this is due to using a square wave for switching at the decoder. A square wave could not be used at the coder because of the bandwidth limitations, and a sine wave is not used in a switching decoder.

Fortunately, it is very easy to eliminate this crosstalk, and two methods will be described.

METHOD (1)

Suppose that the left-hand signal, instead of being L, is $L + xR$ and that the right-hand signal, instead of being R, is $R + xL$, where x is a fraction depending on the magnitude of crosstalk. Suppose that we subtract a signal $x(L+R)$ from both channels.

The output of the L channel is now $L + xR - x(L+R) = L(1-x)$

The output of the R channel is now $R + xL - x(L+R) = R(1-x)$

There is now complete separation of the signals and there is no longer any crosstalk. The magnitude has been reduced slightly, but this is of no importance. A circuit for doing this is given in figure 9.22. The $L+R$ is, of course, the mono signal and available from the demodulator of the receiver.

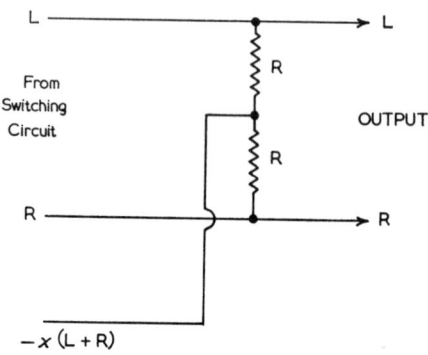

FIG. 9.22.   ONE METHOD OF CANCELLING CROSSTALK

METHOD (2)

The circuit is shown in figure 9.23, where Tr_1 and Tr_2 are the L and R amplifiers with emitter resistors R_1 and R_2. Between the emitters is connected the circuit $R_3 C_1$ which, as we shall see, if correctly adjusted will eliminate the crosstalk. Assuming the input signals from the decoder at L_{in} and R_{in} are $L + xR$ and $R + xL$, then almost the same voltage will be generated across R_1 and R_2 (without $C_1 R_3$) since Tr_1 and Tr_2 are acting as emitter-followers as regards R_1 and R_2. As regards the emitter of Tr_1 there will be a voltage fed through $R_3 C_1$ from Tr_2. The voltage on the emitter of Tr_1 from Tr_2 will be less than that on the emitter of Tr_2 because of R_3 (C_1 is only a d.c. blocking capacitor of low reactance). Suppose that the fraction on the emitter of Tr_1 is p of that on the emitter of Tr_2. Now the voltage on the emitter of Tr_2 is $R + xL$ (or nearly), so the voltage fed to the emitter of Tr_1 will be $p(R+xL)$. The voltage between the base and emitter (which is what controls the collector current) is now $(L+xR) - p(R+xL)$. $L+xR$ is the base to earthy line voltage and $p(R+xL)$ is the emitter to earthy line voltage, and this is subtracted because a voltage fed to the emitter has the opposite effect to the same voltage fed to the base.

Hence the base-emitter voltage is $L + xR - pR - pxL$.

If $p = x$ then this becomes: $L - x^2 L$ or $L(1-x^2)$.

The collector current and output voltage will be proportional to this, and hence the crosstalk has been removed. The same argument applies to the other channel. In practice R_3 is adjusted until the crosstalk is a minimum.

After the switching circuit it is necessary to use a de-emphasis circuit, i.e. an $R - C$ circuit with a time constant of 50 μs. Although there should be little subcarrier from a balanced demodulator the de-emphasis circuit reduces it still further. When a tape recorder is fed from a decoder, if there is any remaining 38 kHz carrier it may cause whistles with the bias oscillator of the recorder. Thus an additional filter is often added to reject any 38 kHz carrier. The circuit may be a parallel-tee filter as in

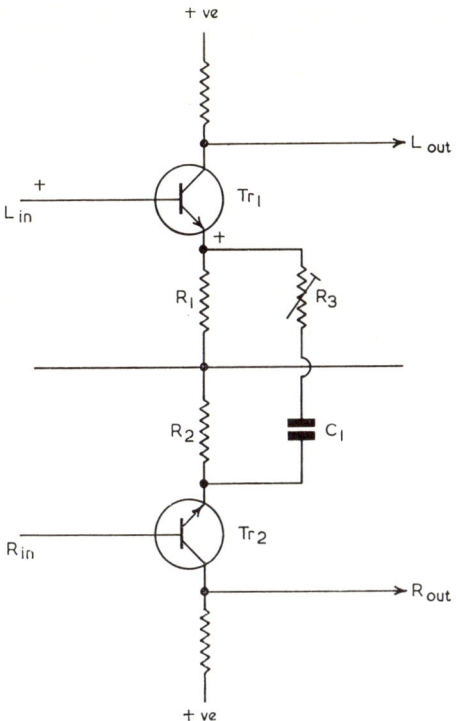

FIG. 9.23. ANOTHER METHOD OF CANCELLING CROSSTALK

figure 9.24. This filter has a high rejection at a frequency given by:

$$f = \frac{1}{2\pi RC}$$

Instead of this circuit, a series LC circuit may be used, as in figure 9.25, the filter being tuned to 38 kHz, the resonant frequency being given by

$$\frac{1}{2\pi\sqrt{LC}}$$

At the resonant frequency the impedance is low, and hence forms nearly a short circuit.

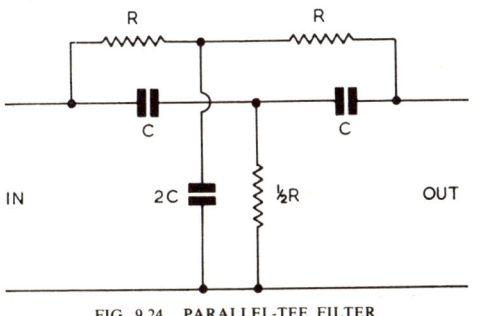

FIG. 9.24. PARALLEL-TEE FILTER

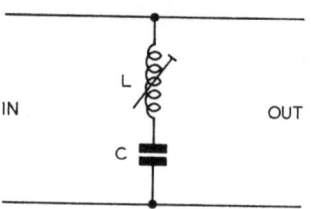

FIG. 9.25. SERIES TUNED CIRCUIT REJECTION FILTER

PRODUCTION OF 38 kHz CARRIER

To operate the subcarrier demodulator or switching circuit it is necessary to have a subcarrier of 38 kHz, which is obtained by frequency doubling from the pilot tone. A typical circuit is given in figure 9.26. The composite signal from the f.m. demodulator is fed into transistor Tr_1 which has suitable bias, but the biasing circuit is not shown. In the collector of Tr_1 is the tuned circuit $C_1 L_1$ and this is tuned to 19 kHz. Thus the only voltage produced across this tuned circuit will be that of the pilot tone of 19 kHz, since the impedance of the circuit will be low at other frequencies. Coupled to the coil L_1 is the centre-tapped coil, which feeds D_1 and D_2. These rectifiers act as full-wave rectifiers and produce a voltage across R_2. The waveform of this voltage is shown in figure 9.27. As shown dotted, the fundamental component of the voltage across R_2 is twice the frequency across $L_1 C_1$. The voltage across R_2 is fed through the d.c. blocking capacitor C_3 to the base of Tr_2. R_3 and R_4 provide suitable bias. The collector of Tr_2 has a tuned circuit $L_5 C_5$ which is tuned to 38 kHz. This then picks out the fundamental component of the voltage across R_2 and rejects the harmonics. Thus a sine wave of 38 kHz is produced across $L_5 C_5$ and a coupling winding

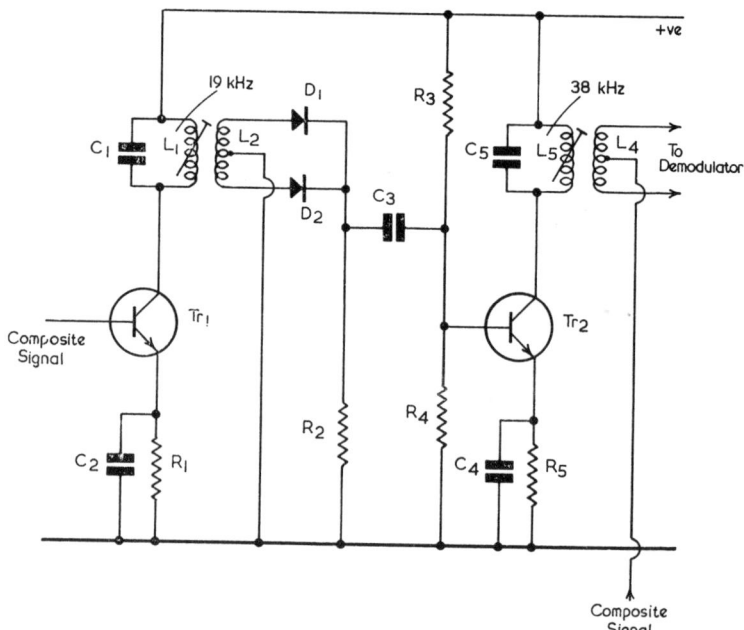

FIG. 9.26. FREQUENCY-DOUBLER CIRCUIT FOR PRODUCTION OF SUBCARRIER

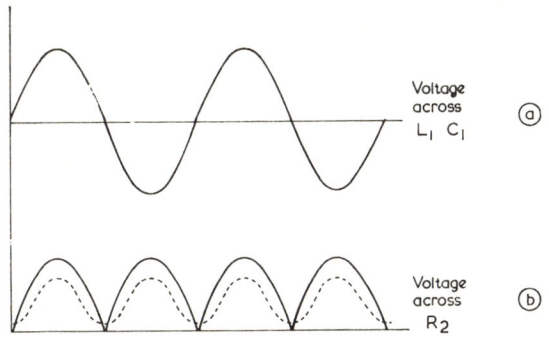

FIG. 9.27. WAVEFORMS OF FREQUENCY-DOUBLER. (a) INPUT; and (b) OUTPUT

L_4 feeds the demodulator, this being as figure 9.19 if it is a switching decoder. Instead of using a full-wave rectifier a class-C amplifier can be used. It is important that the phase of the subcarrier is correct, which may be adjusted by varying L_1 or L_5.

It is common practice to have some form of indication that the signal is a stereo signal by using what is often called a **stereo beacon** or **MPX beacon** (MPX = multiplex). If the transmission is not in stereo then no pilot tone is transmitted, and hence there will be no regenerated subcarrier. The lamp is therefore operated by the presence of either pilot tone or subcarrier. A simple circuit is given in figure 9.28. Tr_1 is fed with the subcarrier signal, for example, from the collector of Tr_2 in figure 9.26. When

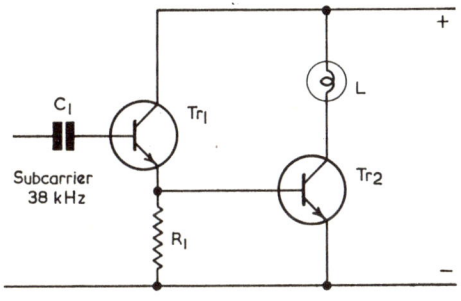

FIG. 9.28. STEREO BEACON CIRCUIT

the subcarrier is present Tr_1 will conduct on positive half-cycles. The resulting voltage across R_1 will turn Tr_2 ON and so operate the lamp L.

A complete circuit of a switching decoder is given in figure 9.29. The composite signal from the f.m. demodulator is fed through C_1 to the base of Tr_1, bias being supplied by R_1 and R_2. In the collector circuit is the tuned circuit C_4L_1, tuned to 19 kHz, which picks out the pilot tone. This is rectified (full wave) by D_1 and D_2 and fed to the base of Tr_2. In the collector is the circuit C_6L_4 tuned to 38 kHz, and so the subcarrier is developed across this circuit and fed to L_5, which feeds the diode switching circuit consisting of D_3 to D_6.

The emitter circuit of Tr_1 contains the series tuned circuit L_3C_3 tuned to 19 kHz. C_2 is a d.c. blocking capacitor. This circuit serves two purposes:

(a) It forms a low impedance at 19 kHz and hence there is little negative feedback and the gain of Tr_1 is increased. This helps in the picking out of the pilot tone by L_1C_4.

(b) It shorts out practically all the pilot tone voltage across R_3 and hence this is not fed to the demodulator or switching circuit.

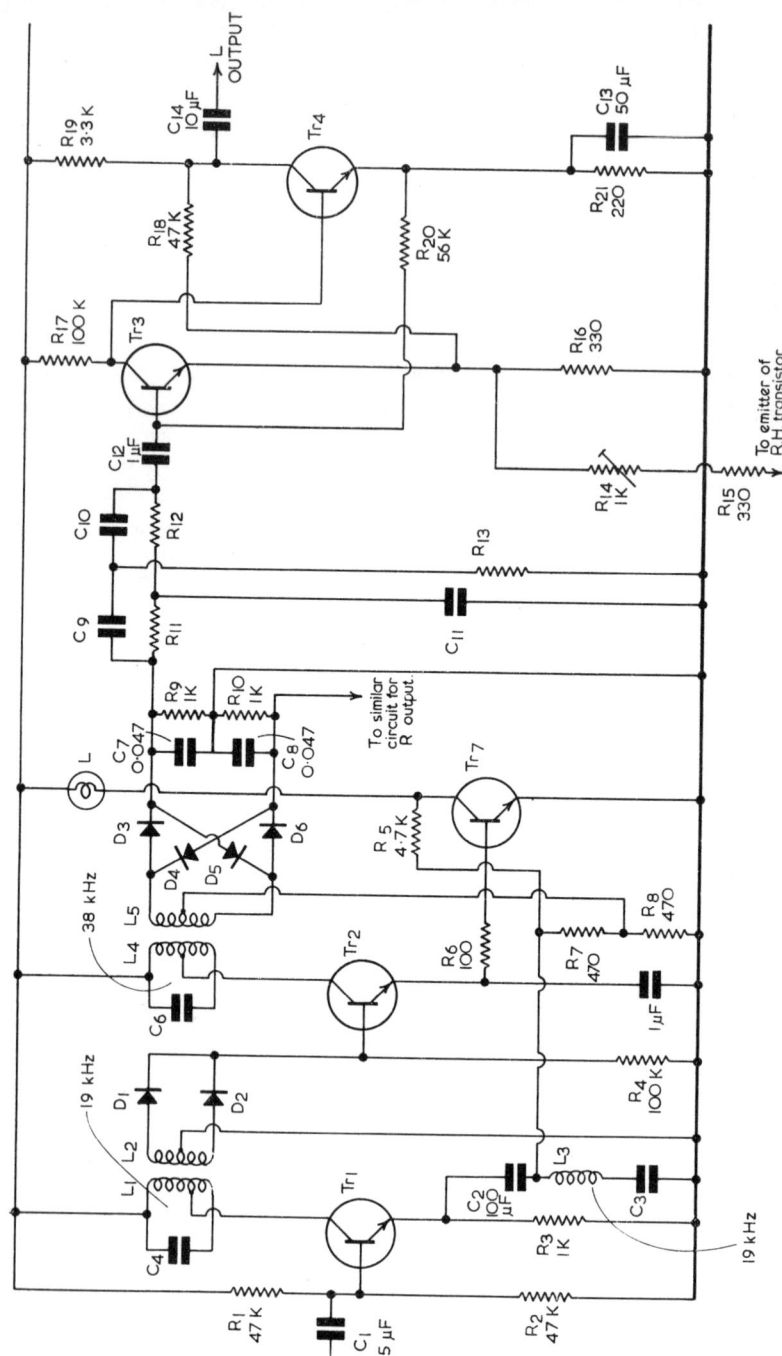

FIG. 9.29.   COMPLETE CIRCUIT OF STEREO DECODER (SWITCHING TYPE)

The composite signal (less pilot tone) across L_3C_3 is fed to the diodes D_3 to D_6 through the potential divider R_7R_8. The resulting L and R signals are developed across R_9 and R_{10}. C_7 and C_8 will remove much of the subcarrier which may be present and also act as the de-emphasis circuit. A parallel-tee filter C_9, C_{10} and C_{11} with R_{11}, R_{12} and R_{13} remove any 38 kHz which is left. A similar circuit is, of course, used in the right-hand channel. Tr_3 and Tr_4 is a two-stage amplifier with feedback through R_{18} and R_{20}, and feeds the main left-hand amplifier. R_{14} and R_{15} are for crosstalk compensation, as described earlier. On a mono transmission Tr_7 will be OFF and the lamp L out. On a stereo signal the current of Tr_2 passes through R_6 to the base of Tr_7 so switching it ON and lighting the lamp L, thereby indicating a stereo transmission. On a mono transmission it is necessary to make two of the diodes conduct, so that the signal fed from across L_3C_3 will pass to both L and R channels. In this circuit this is done by R_5. When the transmission is mono the lamp is out, and hence full positive line voltage is fed through R_5 and R_7R_8 to the diodes. This will turn D_3 and D_6 ON, so completing a path to both channels. When the transmission is stereo the lamp L is lit and Tr_7 is fully ON. Its collector voltage is therefore low and a negligible voltage is fed to the diodes.

Integrated circuits are being used as decoders and two circuits will be described. One is shown in figure 9.30. This operates similarly to the switching decoder. The composite signal from the f.m. demodulator is fed through C_1 to the base of Tr_1, this operating as an emitter-follower. The voltage across the emitter resistor R_1 is fed to two places Tr_2 and Tr_6. Now consider the path through Tr_2. On the base is the tuned circuit L_1C_3 fed through the blocking capacitor C_2. This circuit is tuned to the pilot tone of 19 kHz and so this circuit tends to eliminate other frequencies. In the collector of Tr_2 is a further tuned circuit L_2C_4, this again being tuned to 19 kHz. The voltage across this tuned circuit is fed to Tr_3 and Tr_4, which feed the base of Tr_5. In the collector circuit of Tr_5 is the tuned circuit C_5L_3, one end of which is taken to the d.c. supply from Tr_7, a tap being used to reduce the damping on the circuit. This tuned circuit is tuned to 38 kHz and provides the subcarrier voltage. The collector is directly coupled to the base of Tr_{10} and Tr_{13}. The bases of Tr_{11} and Tr_{12} are returned to the emitter of Tr_7 (and opposite end of tuned circuit) and are therefore at the same d.c. potential. The voltage across R_1 is also fed to Tr_6, which is an emitter-follower feeding the base of Tr_8. Tr_8 and Tr_9 form a kind of long-tailed pair with a common emitter resistor R_2, but the coupling is limited by R_3 and R_4. Therefore, a voltage applied to the base of Tr_8 will result in a collector current in Tr_8 and also collector current in Tr_9 of smaller magnitude. The reason for this is explained later. Tr_{10} and Tr_{11} form a long-tailed pair with Tr_8 in the common-emitter circuit. On one half-cycle of the subcarrier Tr_{10} is ON and Tr_{11} is OFF. Thus the composite signal is fed from Tr_8 and through Tr_{10} to the L output. On the other half-cycle Tr_{10} is OFF and Tr_{11} is ON, so that the composite signal from Tr_8 now passes through Tr_{11} to the R output. Thus Tr_{10} and Tr_{11} act as a switch in a similar way to the four diodes of figure 9.19. In the same way Tr_{12} and Tr_{13} form a long-tailed pair with Tr_9 in the common-emitter lead. Tr_9 feeds the opposite polarity signal to the L and R outputs through Tr_{12} and Tr_{13}. When Tr_{10} is ON so is Tr_{13}, so a positive signal is fed through Tr_{10} to the L output and a negative signal is fed through Tr_{13} to the R output. Similarly, when Tr_{11} and Tr_{12} are ON a positive signal is fed through Tr_{11} (from Tr_8) to the R output and a negative signal (from Tr_9) is fed through Tr_{12} to the L output. Thus each output is fed with the normal signal which, as we have seen, will have some crosstalk. Let the L output due to Tr_{10} and Tr_{11} be $L + xR$. The L output is also fed with a reversed signal of $p(R + xL)$ where p depends on the values of R_2, R_3 and R_4. The overall signal is therefore $L + xR - p(R + xL)$ and if p is made equal to x, then this becomes $L - x^2L$ or $L(1 - x^2)$, i.e. the crosstalk is removed. Tr_9, Tr_{12} and Tr_{13} are used to eliminate crosstalk from both outputs in this way. When the transmission is stereo the emitter current of Tr_5 turns Tr_{18} ON, which turns Tr_{14}, Tr_{15}, Tr_{16} and Tr_{17} ON, so lighting the stereo lamp L. The remainder of the circuit is used to provide suitable bias voltages to the transistors.

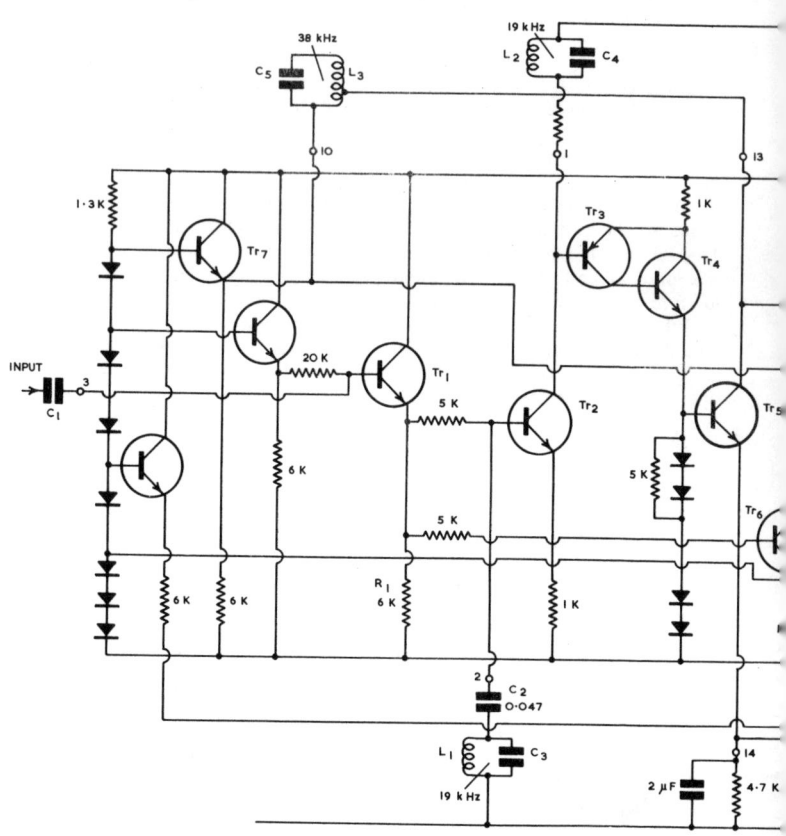

FIG. 9.30. INTEGRATED CIR

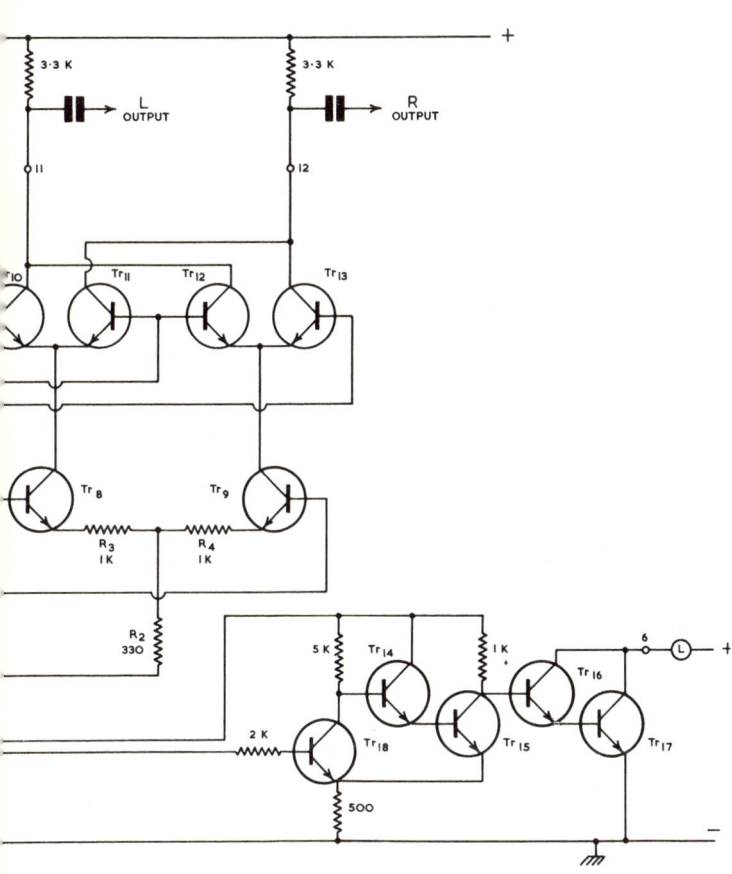

HING DECODER (MC 1307)

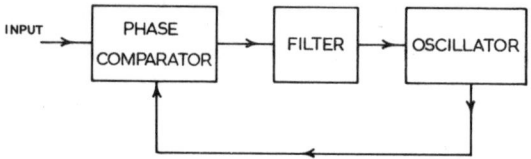

FIG. 9.31. PRINCIPLE OF PHASE-LOCKED LOOP

The second type of integrated circuit operates on quite a different principle to those described, and makes us of what is called a "phase locked loop". The principle of the phase locked loop is given in figure 9.31. The idea is to lock an oscillator to the incoming frequency, in this particular case the pilot tone frequency. The input is fed to a phase comparator which compares the phase of the input waveform with that from the oscillator. An error signal is produced which, after filtering, is fed to the oscillator to control its frequency. Thus, within limits, the oscillator will follow the frequency of the input. Other frequencies on the input, provided they are reasonably different to that of the oscillator, will have no effect, *i.e.* the device also acts as a filter to pick out a particular frequency, the frequency being determined by the nominal frequency of the oscillator. A block diagram of the decoder is given in figure 9.32. The input from the f.m. demodulator is fed through a buffer stage and then takes three paths. One is to the phase comparator or phase lock detector (sometimes called a multiplier). The oscillator does not run at the pilot tone frequency but at a frequency of four times the pilot tone frequency (*i.e.* 76 kHz). The oscillator frequency is divided twice by two, and the resulting 19 kHz fed to the phase comparator. The error signal is fed to a low-pass filter (to cut out higher frequencies) to a d.c. amplifier which controls the frequency of the oscillator. The low-pass filter determines the characteristics, such as pull-in range and control range. This phase locked loop produces the 38 kHz subcarrier. The output of the oscillator, after going through one $\div$ 2 circuit and stereo ON-OFF switch, is fed to the stereo demodulator or switching circuit, which is also fed with the composite signal. The L and R outputs are fed through de-emphasis circuits to the L and R output connections. The remainder of the circuit is concerned with the stereo beacon. The oscillator output, after dividing twice by two, is fed to the pilot tone detector, which detects whether the oscillator is in phase with the pilot tone, *i.e.* which it will be if it is a stereo transmission. The output is fed to the Schmitt trigger (a transistor switch circuit) through the low-pass filter. The Schmitt trigger operates the lamp driver and switches ON the stereo beacon on a stereo transmission. At the same time it closes the stereo ON-OFF switch, so feeding subcarrier to the stereo demodulator block. If there is no pilot tone then the Schmitt trigger is not operated, and the beacon lamp is OFF and no subcarrier is fed to the stereo demodulator (the oscillator still runs but not at the correct frequency). It is not intended to deal with the circuits as they differ in detail between different integrated circuits. The phase lock detector is commonly a switching circuit to $Tr_8 - Tr_{13}$ of figure 9.30. This phase detector can be considered as equivalent to a multiplier. The voltage-controlled oscillator is an astable multivibrator, whose frequency can be controlled by the voltage. The pilot tone detector and stereo demodulator again are similar circuits to Tr_8-Tr_{13} of figure 9.30. One may wonder why an oscillator of 76 kHz is used instead of 38 kHz. It is difficult to make an astable multivibrator which has a square wave output which is symmetrical (*i.e.* equal mark-to-space ratio) particularly when controlled by a voltage (this type of oscillator being known as a voltage controlled oscillator or VCO). By using an oscillator of twice the subcarrier frequency it makes sure that the 38 kHz switching waveform is symmetrical. This is essential if correct switching is to occur at the stereo demodulator. The reason why the 38 kHz is symmetrical is shown in figure 9.33, where it will be seen that any asymmetry in the waveform has no effect on the symmetry of the 38 kHz waveform because each half-cycle corresponds to one cycle of the oscillator waveform. This type of decoder has the advantage that it has no tuned circuit and is therefore very easily set up. The only

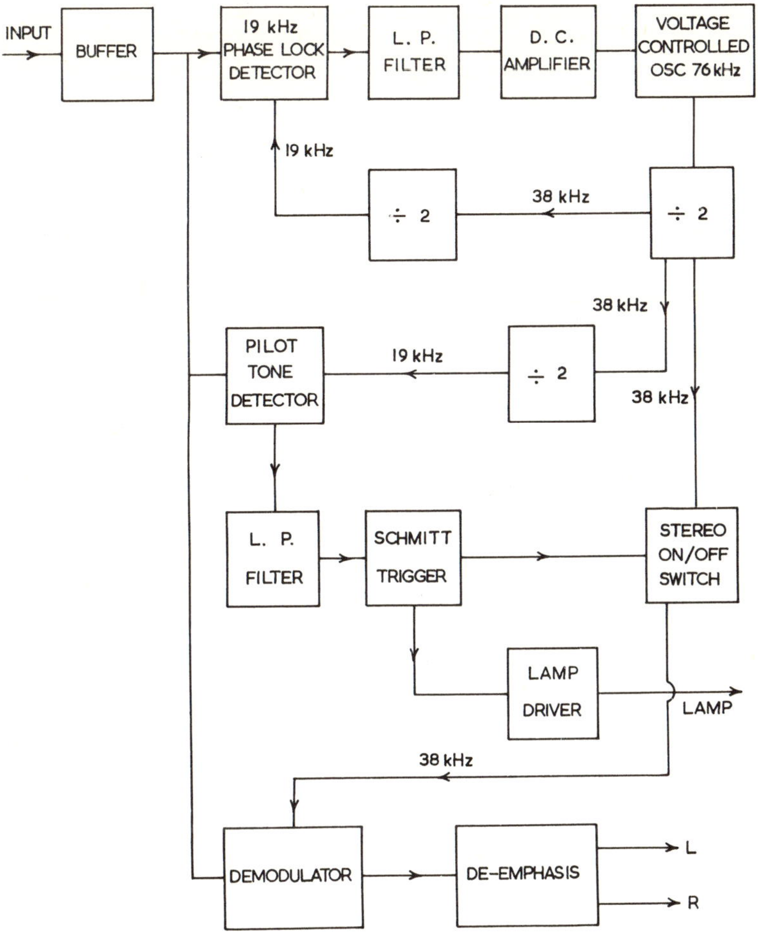

```
INPUT → [BUFFER] → [19 kHz PHASE LOCK DETECTOR] → [L. P. FILTER] → [D. C. AMPLIFIER] → [VOLTAGE CONTROLLED OSC 76 kHz]
```

19 kHz

[÷ 2] ← 38 kHz ← [÷ 2]

[PILOT TONE DETECTOR] ← 19 kHz ← [÷ 2] ← 38 kHz

38 kHz

[L. P. FILTER] → [SCHMITT TRIGGER] → [STEREO ON/OFF SWITCH]

[LAMP DRIVER] → LAMP

38 kHz

[DEMODULATOR] → [DE-EMPHASIS] → L → R

FIG. 9.32. INTEGRATED CIRCUIT DECODER USING PHASE-LOCKED LOOP

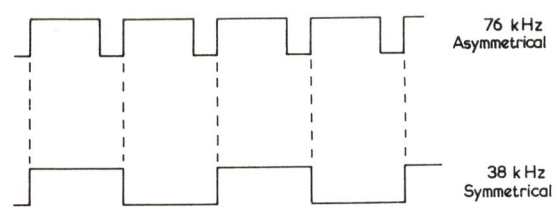

76 kHz Asymmetrical

38 kHz Symmetrical

FIG. 9.33. REASON FOR USING 76 kHz OSCILLATOR

preset adjustment might be a resistor in the oscillator circuit to set the frequency to its correct nominal value. It operates with a small input and has low distortion.

Other decoder circuits are possible, but these are not being included as they are not commonly used.

NOTE ON WAVEFORMS

If the waveform from the f.m. demodulator of a receiver is viewed on a cathode ray oscilloscope the pilot tone (and subcarrier) can be seen if the transmission is in stereo. However, if a mono transmission is viewed it will be found to look very similar, there being a high frequency of constant amplitude. This is a pilot frequency used by the BBC for test purposes and has a frequency of 23·3 kHz. This, of course, does not operate the decoder because of the frequency difference.

STEREO AMPLIFIERS

In a stereo system two identical amplifiers are required, preferably feeding two identical speakers. For simplicity the two volume controls are usually ganged. If tone controls are fitted these too are usually ganged as, in general, one should alter the response of both amplifiers by the same amount. Due to the fact that the amplifiers may not have exactly the same gain, it is usual to fit a balance control which varies the gain of one amplifier relative to the other. The simplest way is a volume control in each channel, as shown in figure 9.34, and so arranged that as one potentiometer increases the gain the other one reduces it, *i.e.* one is connected the reverse way.

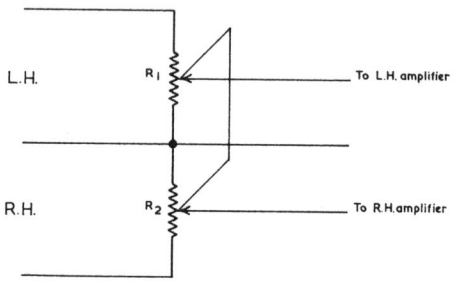

FIG. 9.34. STEREO BALANCE CONTROL

Because one is reversed they should be linear potentiometers and not logarithmic. The disadvantage of this arrangement is that if normal potentiometers are used then, in the centre position, the gain of both amplifiers is reduced, because both potentiometers are in their mid position. One way of overcoming this is to use special potentiometers where the upper half of each track is fully conducting. In this way full gain on both channels is obtained in the centre position. If the control is turned in one direction the gain of one amplifier will remain constant, while that of the other is reduced. Another method is given in figure 9.35, where a special centre-tapped potentiometer is used. When the control is in the centre position the gain of each channel will be at a maximum. If it is moved in one direction the gain of one amplifier remains constant, but the other is reduced because the value of the resistor on the base is reduced. The gain is reduced in these circumstances because of the voltage drop due to the output impedance of the previous stage. Another alternative is given in figure 9.36. When the control is moved from the centre position the resistance in the base circuit of one amplifier is increased, while that in the other is reduced. Thus the gain of one amplifier increases while that of the other decreases. This circuit has the advantage of not requiring a special potentiometer which should be linear. Another method is given in figure 9.37. Resistors R_1 and R_2 are the normal emitter resistors which are unby-passed. $C_1 = C_2$ and have such values that their reactances are low. If P_1 is in the central position (assuming a linear potentiometer) there will be considerable resistance in series with each capacitor, and the emitter resistors are virtually unby-passed and the gain is reduced by negative feedback across R_1 and R_2. If P_1 is varied then the resistance in one side will be reduced, hence the emitter re-

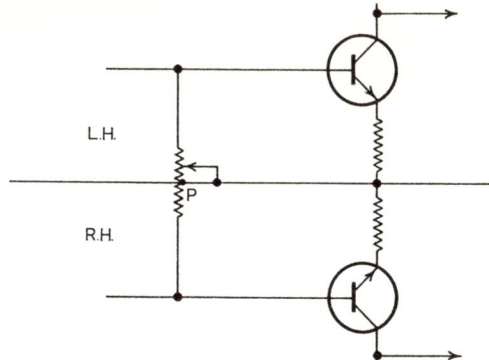

FIG. 9.35. ALTERNATIVE BALANCE CONTROL

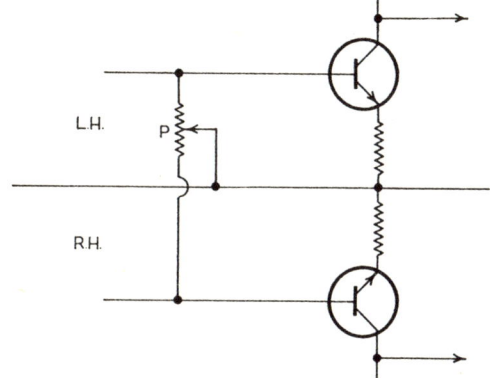

FIG. 9.36. BALANCE CONTROL SIMILAR TO FIGURE 9.35

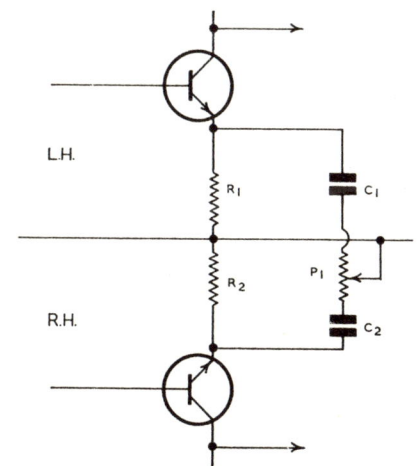

FIG. 9.37. BALANCE CONTROL USING NEGATIVE FEEDBACK

sistance is to some extent by-passed and the gain will increase. In the other channel the resistance is increased, tending to reduce the gain.

One method of solving the problem of the balance control is to fit separate volume controls. If these are made slider controls side-by-side it is easy to move them together for normal changes of volume or move one relative to the other to obtain the correct balance.

PHASING OF LOUDSPEAKERS AND ADJUSTMENT OF BALANCE CONTROL

It is possible for the two speakers to be connected so that, on mono, the sounds do not add but subtract. In these conditions the stereo reproduction is not correct. One method of checking this is to play a mono record or tape through both amplifiers. (Many amplifiers have a mono-stereo switch which parallels the two amplifier inputs in the mono position: in this case the signal source can be stereo). If the phasing is correct the sound should appear to come from a position between the two speakers. If this is not the case then the leads to ONE speaker should be reversed. If the sound comes from a point between the two speakers but not from the centre position then the balance control should be adjusted until the sound appears to come from the centre.

An alternative method of checking is to place the speakers close together, preferably facing each other. The amplifier may be fed with music (preferably mono, or the amplifier switched to the mono position) or from an audio frequency oscillator. If music is used it should be chosen to have plenty of bass. If the phasing is correct then the bass should sound the same whether they are close together or apart. If the bass is much reduced when they are placed together then the phasing is incorrect. If an oscillator is used it should be set to a low frequency, say 50–100 Hz. Instead of moving the speakers they may be placed close together and one disconnected. If, on disconnecting one of the speakers the volume of the bass increases, then they are incorrectly phased and one should be reversed.

CHAPTER 10

RECORD REPRODUCTION

PICK-UPS

THE purpose of a pick-up is to convert the variations in the groove of a record into corresponding voltage variations. This might sound simple but in practice it is complex, particularly if high-quality results are to be obtained and record wear is to be small. The introduction of stereo records increased the complexity.

We will first consider a mono record and the nature of the recording process. The master record is made by using a cutter driven by a lead screw across the record, and the cutter is made to oscillate sideways in proportion to the magnitude of the signal being recorded. The waveform of the signal will normally be complex and continually changing, but for simplicity we will assume a sinusoidal signal. At a constant frequency the groove will look like figure 10.1. The magnitude of the side-to-side movement A will depend on the amplitude of the signal, but the maximum

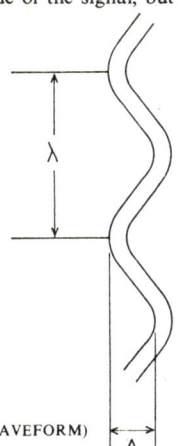

FIG. 10.1. RECORD GROOVE (SINE WAVEFORM)

movement must be limited or the grooves will run into each other. Obviously the maximum allowable movement will depend on the spacing of the grooves. This spacing is different from record to record and may be made variable on a single record, the spacing being increased on loud passages. The distance between two peaks (in the same direction) on the record groove (λ on the figure) is called the wavelength. This will vary with the frequency of the recorded signal and on the speed of the record. The relationship between wavelength λ, the frequency f and the speed s of the groove is given by

$$\lambda = \frac{s}{f} \qquad\qquad\qquad \textbf{(10.1)}$$

where λ = wavelength in metres

s = groove speed in metres/sec

f = frequency in Hertz.

(This is a similar expression to that connecting frequency and wavelength of radio waves).

The speed of the groove will depend on:

(a) the record speed in r.p.m.; (b) the part of the record being played.

(a) The speed of the record is a compromise between good reproduction and playing time. Originally records running at 78 r.p.m. were used, these being available in diameters of 10″ and 12″. The material used for these records did not allow fine grooves to be moulded and only about 40 grooves per cm (100 grooves/inch) were used. The material was granular and caused a lot of noise. These are now obsolete. With the development of new plastic materials it was possible to use finer grooves, and hence get more playing time. The material also allowed finer groove variations to be moulded so that the speed could be reduced. These long-play (or microgroove) records use about 100 grooves per cm (250 grooves/inch). Two speeds are used: 45 r.p.m. on 7″ records and 33 r.p.m. with 10″ and 12″ records (some 7″ records are also available for this speed). 12″ records are much more common than 10″.

(b) The speed will be a maximum at the outer edge and a minimum on the inside groove. To give some idea of the wavelength, consider the outer groove of a 12″ (30 cm) record with a frequency of 5 kHz. The speed is $2\pi r \times n$ where r is the radius in metres and n is the speed in revolutions per second. In this case

$$r = \frac{15}{100} \text{ metres and } n = \frac{33}{60}.$$

The speed is therefore $2\pi \frac{15}{100} \times \frac{33}{60}$ m/s $= 0.52$ m/s.

Substituting this in equation (**10.1**)

$$\lambda = \frac{0.52}{5000} = 0.0001 \text{ m} = 0.01 \text{ cm (about } \frac{4}{1000} \text{ inch)}$$

At the centre λ will be smaller, the radius being about 6 cm (depending on the record). The wavelength now works out at 0.004 cm. Thus the side-to-side movement takes place in a very short distance. As the frequency is decreased the wavelength increases.

As the stylus moves in the groove it is subject to rapid changes in velocity as it follows the sinusoidal displacement of the groove. At point A of figure 10.2 the stylus has moved a maximum amount in one direction and is stationary at this instant,

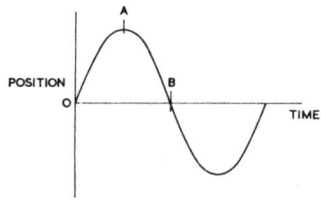

FIG. 10.2. SINE WAVEFORM. MAXIMUM AMPLITUDE AT A AND MAXIMUM VELOCITY AT B

i.e. the velocity is zero. At point B the stylus is in its centre position but moving with maximum velocity. The greater the amplitude of the groove modulations the greater will be the velocity, the velocity being proportional to the amplitude of the groove modulation. As the frequency is increased the time for one cycle is reduced, and hence for a given amplitude of modulation the greater will be the velocity, in fact, the velocity is proportional to the frequency. The maximum velocity is given by $V_m = 2\pi f A$ when A is the amplitude of the groove modulations.

The range of frequencies which we wish to record is high, say, at least 100 Hz to 10,000 Hz. This is a ratio of 100/1. If the record were recorded with constant amplitude of groove modulations then the velocity would change over the same range and be very low at low frequencies and extremely high at high frequencies; it would be difficult to get a stylus to follow the grooves. Alternatively, we might use constant velocity recording, but if this were done the amplitude of modulation of the grooves would be very small at high frequencies but large at low frequencies, tending to result in large groove spacing.

In order to get over these problems a combination of the two is used and records are now made to a standard known as RIAA (Radio Industry Association of America).

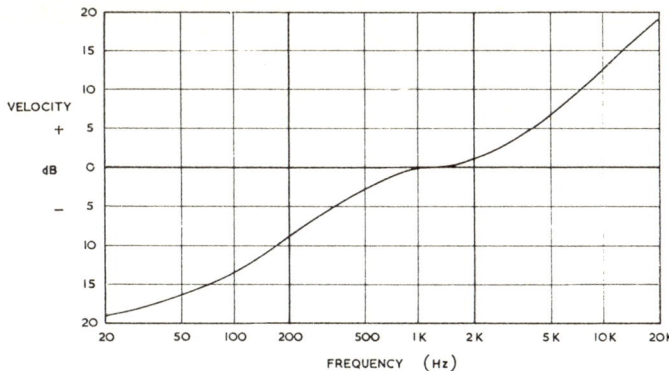

FIG. 10.3. R.I.A.A. RECORDING CHARACTERISTIC

This standard is illustrated in figure 10.3. This is the normal way of showing the characteristic, but some explanation is necessary. The horizontal scale is straightforward and is the frequency plotted on a logarithmic scale. The vertical scale is shown in dB. This is intended to represent the output that would be obtained from a pick-up which had an output proportional to velocity. Hence the scale can be marked off in velocity, although it is often just quoted in dB. Around 1 kHz there is a region of constant velocity recording (constant velocity corresponds to a horizontal line and the amplitude is then inversely proportional to the frequency). Above this frequency the amplitude of the groove variations is approximately constant and therefore the velocity is approximately proportional to frequency. For frequencies below the constant velocity section, the amplitude is maintained constant and hence the velocity decreases. Below about 100 Hz the rate of decrease in velocity is reduced, the amplitude increasing slightly. In a general way one may think of the recording as being one at constant amplitude, but by having the centre portion at constant velocity the change of velocity between the lowest and highest frequencies is reduced. The maximum velocities recorded may be as high as 25 cm/second or 0·25 m/s.

Since the velocity of the stylus is changing rapidly it is subject to rapid acceleration and deceleration. Since any stylus has mass it can only be accelerated or decelerated by applying a force to it, and the greater its mass the greater the force that must be applied, the force also being proportional to the acceleration or deceleration. As well as the actual mass of the stylus there is also what is called the "reflected mass", due to the mass of the remainder of the cantilever, etc. The total mass, *i.e.* actual plus reflected, is called the "equivalent" or "effective" mass, and it is this figure that matters. The force required to accelerate or decelerate the stylus must, of course, come from the walls of the groove, and this is shown in figure 10.4. The side walls are at 45° to the vertical, and the force F they apply will be at right angles to the walls, as shown.

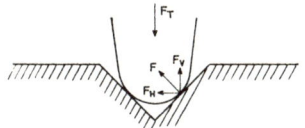

FIG. 10.4. FORCES ON STYLUS IN GROOVE

This can be considered as equivalent to two forces: a horizontal force F_H and a vertical force F_V. The force F_H will accelerate or decelerate the stylus, but force F_V will try to lift the stylus out of the groove. Therefore, to keep the stylus in the groove we must apply a downward force F_T which at all times must be greater than F_V so that the stylus is kept in contact with the groove walls. However, the greater this force F_T the greater the force on the walls. The contact area of the stylus is small, and hence large pressures are produced, and if too great the walls will be deformed and distortion will occur. The deformation may be permanent if the pressure is too high and the record is then damaged. Thus modern pick-ups are designed to have a small stylus effective mass, say less than 1 milligram. As a result the downward force F_T is small, say $1\frac{1}{2}$ milligrams, but may be up to 10 milligrams in cheap pick-ups.

Another factor also comes in related to the tracking of the pick-up. The stylus must be held in the correct central position by the equivalent of a spring. The device that does this has a certain stiffness, *i.e.* it requires a certain force F to move it a distance A from its central position. The stiffness is defined as F/A. However, the inverse of this A/F is normally quoted and called the "compliance", *i.e.* it is the amount the stylus moves for a given applied force. The unit of force commonly used is the dyne, and there are approximately 1000 dynes in a gram (not really correct but satisfactory for the present purpose). The unit used for the movement is the centimetre, although the movement is very small. Thus the compliance is the amount that the stylus moves for a force of 1 dyne applied to it. The larger the compliance the better because it means, for a given movement, a less force has to be applied to the stylus by the groove walls. Thus, the force produced by the groove walls must be sufficient to produce the acceleration or deceleration, and also overcome the stiffness of the stylus. A figure of 1 to 5×10^{-6} is typical of the compliance for cheap pick-ups, but for high grade pick-ups this increases to 20 to 30×10^{-6} cm/dyne.

As well as the force required to move the stylus back and forwards there must be a force that moves the pick-up towards the centre of the record. If the pick-up arm does not move freely then, with a pick-up having a large compliance, the stylus may be displaced appreciably from the central position and may, in fact, be damaged. For the same reason, a high compliance pick-up is more easily damaged by careless use. Thus, when using high-grade pick-ups with a high compliance, a good arm which moves very freely must be used. Pick-up arms are considered later in this chapter. It is also important that resonances should not occur in the arm; if there are any they should be well damped, otherwise bad tracking and corresponding distortion will result.

Having dealt with mono recordings (most recordings are now stereo) we will consider stereo recordings. We now have to record two signals, the left (L) and right (R), using the same groove. In a mono record the modulation of the groove is horizontal or lateral. In very early records the modulation was up and down or vertical, and known as "hill and dale" recording. In a stereo record we use both at the same time. It might appear that we could use say horizontal modulation for the L signal and vertical modulation for the R signal. If this were done the record would not be compatible, *i.e.* it could not be played satisfactorily on a mono record player. If it was played on a mono record player then the output would correspond to the L signal only (the horizontal modulation) since the pick-up would not produce an output for the vertical modulation—obviously unsatisfactory. Accordingly, the horizontal modulation corresponds to the sum signal L + R (which is the same as mono) and the vertical modulation to the difference signal L − R. This now results in a compatible record.

Due to the fact that the walls are at 45° to the vertical one wall can only move the stylus in a direction at 45° to the vertical. One wall is modulated with the L signal and the other with the R signal, the directions of the movements being shown in figure 10.5. Looking at the groove, the position under various conditions is given in figure 10.6. At (a) we have only an L signal and the modulation is applied to the right-hand wall as shown. The pick-up therefore moves in the direction shown by the arrows. There is a horizontal component of movement and a vertical component, as is required, since in these conditions the sum signal is L and the difference signal is L(R = 0). At (b) is shown the conditions when there is only an R signal. Consider now where R = L and they are in phase, or R = +1 and L = +1. The right-hand wall is modulated with the L signal and the left-hand wall with the R signal; but when the left-hand wall goes up, the right-hand wall goes down by the same amount. Thus the

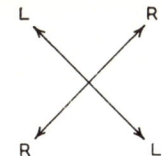

FIG. 10.5. DIRECTIONS OF MODULATION OF STEREO RECORD

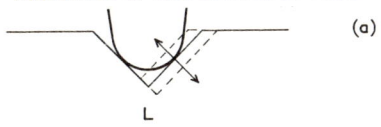

(a)

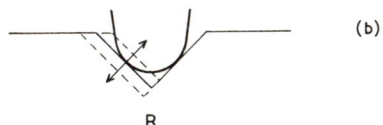

(b)

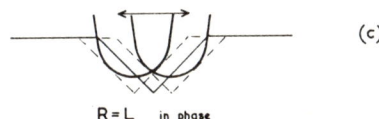

(c)

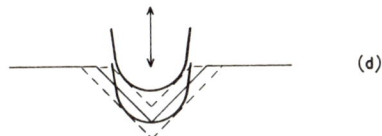

(d)

R = L 180° out of phase

FIG. 10.6. STEREO RECORD GROOVE

stylus moves from side to side as shown. In this condition the sum signal is $L + R = 2L$ (since $L = R$) and the difference signal is $L - R = 0$. Thus there is horizontal movement and no vertical movement. At (d) is shown the position when L and R are equal, but of opposite phase, *i.e.* $L = -R$. In this case the right-hand wall is modulated with the L signal and the left-hand wall with the R signal, but the walls go up and down together. Thus the stylus moves up and down as shown, there being no horizontal movement. In this case the sum signal $(L + R)$ is $L - R = 0$ (when $L = R$) and the difference signal $(L - R)$ is $L - (-R) = L + R = 2L$. Thus there is vertical motion only.

PICK-UPS: PRINCIPLE AND BASIC CONSTRUCTION

We will now consider the principle and basic construction of pick-ups starting with mono pick-ups for simplicity. There are several types:

(a) Moving iron;   (c) Moving coil;

(b) Moving magnet;   (d) Ceramic or crystal.

The first three operate on magnetic principles while the fourth (d) operates on the piezoelectric principle. The basic principle of the moving-iron type is shown in

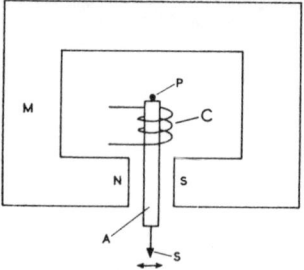

FIG. 10.7. PRINCIPLE OF MOVING-IRON PICK-UP

figure 10.7, where M is a permanent magnet with pole-pieces N and S. Between the pole-pieces is a small iron armature A, pivoted at P to which the stylus S is attached. Around the armature is a coil C, which is not in contact with the armature and hence fixed. Some means must be added (not shown) to maintain A centrally, and this device controls the compliance. When the stylus is in the groove it will be moved from side to side as shown. When A is in the central position the device is symmetrical, and hence no flux will pass through the coil. When the armature is displaced the symmetry is upset and some flux will flow in A and through the coil C. When the armature is displaced in the other direction, the direction of the flux in the coil C will reverse. This changing flux in the coil induces an e.m.f. in it, but it must be remembered that this e.m.f. is proportional to the **rate of change of flux** and this is proportional to the velocity of the stylus and NOT proportional to the displacement. The importance of this will be seen later. The actual mechanical layout may be considerably different to that shown, but the principles remain the same.

Figure 10.8 shows the principle of the moving-magnet type of pick-up. This consists of an iron circuit B with small magnet M pivoted at P and carrying the stylus S.

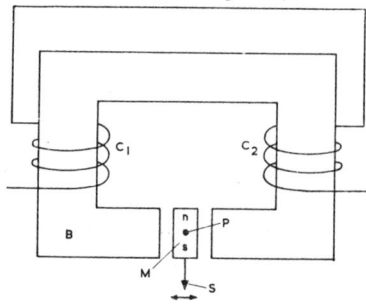

FIG. 10.8. PRINCIPLE OF MOVING-MAGNET PICK-UP

Around the iron circuit are two coils C_1 and C_2. When the magnet is not displaced the device is symmetrical and no flux will flow through the coils. When the magnet is displaced, by movement of the stylus, the magnet is tipped and some flux will flow through the iron system M, and link with the coils. As the magnet moves, this field varies and produces a voltage in the coils. As in the last case the e.m.f. produced will be proportional to the stylus velocity. Again, there are many different physical layouts.

The principle of the moving-coil pick-up is given in figure 10.9. There is a permanent magnet M with poles n and s with a fixed central-iron core A. The coil moves in the air-gap between M and A, and the stylus is fixed to the coil. Thus, as the stylus moves from side to side the coil moves in the gap and an e.m.f. is generated proportional to its velocity. Thus the e.m.f. is proportional to the velocity of the stylus as in the last two cases. In order to keep the mass down, the number of turns that can be used on the coil are small (compared with the other cases where the coil is fixed), and hence the generated e.m.f. is very small. For this reason the coil is normally connected to a

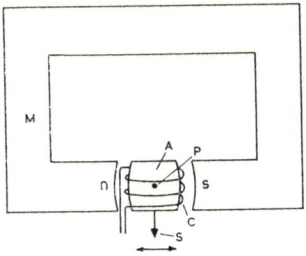

FIG. 10.9.   PRINCIPLE OF MOVING-COIL PICK-UP

step-up transformer, situated near the pick-up and the secondary voltage used to supply the amplifier.

The principle of operation of the ceramic or crystal pick-up is quite different, as it depends on the piezoelectric property. When a piezoelectric material is subject to strain an e.m.f. is generated (between say, metal plates attached to it) and this e.m.f. is proportional to the strain. The basic principle is shown in figure 10.10 where C is the piezoelectric element, which is held in a flexible mount M at the far end. The stylus

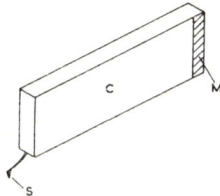

FIG. 10.10.   PRINCIPLE OF CRYSTAL OR CERAMIC PICK-UP

is attached to the front end. As the stylus moves it puts a strain on the element (bending it) and an e.m.f. is generated. Unlike magnetic pick-ups this e.m.f. is proportional to the displacement of the stylus and NOT the velocity.

Turning now to stereo records, it is important that these records can be played on a mono record player (to produce a mono output). On a mono record there is no up-and-down movement of the stylus, and so the compliance in this direction need not be low. In fact, older pick-ups were constructed with a low vertical compliance, and if these are used to play stereo records the records are likely to be damaged due to the large pressures resulting from the pick-up trying to prevent this vertical motion. Modern mono pick-ups are made with a high compliance in the vertical direction as well as in the horizontal direction, so that stereo records are not damaged. It is therefore important to make sure that a mono pick-up is suitable before playing a stereo record.

Turning now to stereo pick-ups, only a few will be described as the number of physical arrangements are numerous. Basically, two pick-up systems are required at right angles to each other. One design of moving-magnet pick-up is shown in figure 10.11. The stylus S is attached to the small moving magnet M by means of a cantilever L. The magnet is held in a resilient material P, which acts as a pivot and also controls the compliance of the system. There are two sets of pole-pieces $P_1 P_2$ and $P_3 P_4$, arranged at $90°$ to each other but $45°$ to the vertical. The pole-pieces form part of an iron system I with coils C, as shown at (b). If the magnet moves along the line $P_1 P_2$ (due to, say, an L signal only) then an e.m.f. will be induced in the coils on the $P_1 P_2$ iron circuit. Since the magnet is moving parallel to P_3 and P_4 little or no e.m.f. will be induced in the coils associated with P_3 and P_4. If only the R signal is present then the movement is along the line $P_3 P_4$ and an e.m.f. is induced in the coils in this system and not in the others. The same idea can be used to produce a moving-iron pick-up. The magnet M is now replaced by, say, a small cylinder of mumetal.

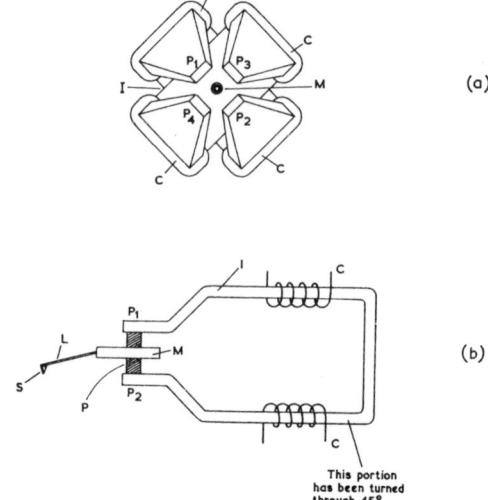

FIG. 10.11. MOVING-MAGNET STEREO PICK-UP

A flux is now produced by incorporating a permanent magnet in the iron system I. The movement of cylinder will cause variations in flux as already explained.

Another moving-iron system (B & O) is shown in figure 10.12. In this case a small iron cross M is used and the stylus S is attached to it by a cantilever C. The

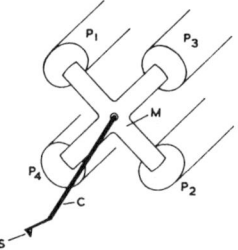

FIG. 10.12. MOVING-IRON STEREO PICK-UP (B & O)

stylus and cross are held in place by some resilient material at the back of the cross. Poles P_1 and P_2 are joined to one permanent magnet with a coil round it and P_3 and P_4 to another magnet and coil system. As the stylus moves it rocks the cross and changes the flux and so induces an e.m.f. in the coils. If only one signal is present then the cross is rocked so that movement takes place between the cross and, say, P_1 and P_2, inducing an e.m.f. in the coils associated with this magnet system. Little movement will take place between the cross and P_3 and P_4, and little e.m.f. is induced in the other coils.

There is one pick-up (Decca) that does not operate in this way and directly produces $L + R$ and $L - R$ signals. For this reason it is sometimes called a sum-and-difference pick-up. The principle is shown in figure 10.13 where the stylus is directly connected to the small magnetized armature. The armature is held in place by the flexible beam and, to prevent undue compliance, the end of the armature is attached to the tie-back thread. If the armature moves side to side (lateral movement) an e.m.f. will be induced in the lateral coil. However, there will be little movement

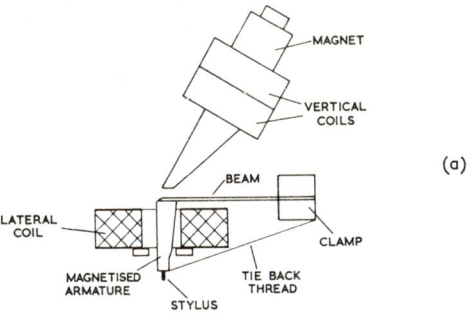

(a)

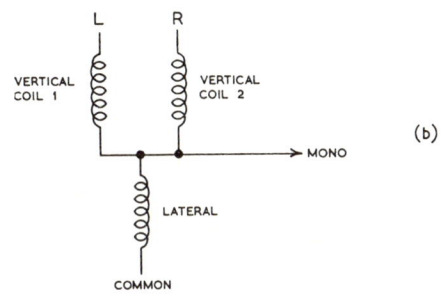

(b)

FIG. 10.13. SUM-AND-DIFFERENCE PICK-UP (DECCA)

relative to the vertical coils and pole-piece, and no e.m.f. is induced in these coils. For vertical movement the gap on the vertical pole-piece will change and an e.m.f. is induced in the vertical coils. However, little voltage will be induced in the lateral coil. The coils are connected as shown at (b). In the case of a mono record or mono reproduction from a stereo record, the output is taken from across the lateral coil. Suppose that there is only groove modulation due to the L signal. In this case the stylus will move equal amounts laterally and vertically. It is arranged that equal voltages are now induced in the lateral coil and in the vertical coil 1 and in such a direction that the e.m.f.s add, so producing the L output. There will be an equal e.m.f. in vertical coil 2, but this is arranged to oppose the e.m.f. of the lateral coil, and hence no signal is produced at the R output. The reverse action takes place for an R signal.

The basic arrangement of a ceramic stereo pick-up is shown in figure 10.14. This consists of diamond shaped plastic bridge *PRQS* used to couple the stylus, attached at *S* to the *L* and *R* ceramic elements, which are clamped at the far ends. The corner *R* is fixed rigidly. If the stylus moves in the *L* direction it bends the diamond piece in the manner shown in figure 10.15, the movement being exaggerated. The plastic hinges at the corners *P* and *R*, and therefore there is little movement of the *L* element. However, the movement of the arm *RQ* causes bending of the *L* element so producing an output. When the stylus moves in the *R* direction a similar action takes place, but an e.m.f. is generated in the *R* element.

The magnetic pick-up, particularly the expensive ones, generally produce better results than the ceramic. The ceramic pick-up is cheap and satisfactory for normal use, and is used in many popular record players. It may be made in the form of a turnover cartridge with two styli, one for 78 r.p.m. records and the other for long play

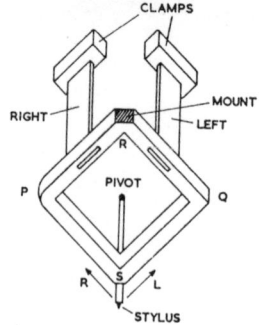

FIG. 10.14. CERAMIC STEREO PICK-UP

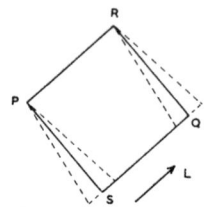

FIG. 10.15. MOVEMENT OF DIAMOND OF STEREO PICK-UP

records (33 and 45 r.p.m.). Two are necessary as the radii of the stylus tips are required
to be different.

STYLUS

For general use a spherical stylus is used, but some distortion will result owing to
the size of the stylus relative to the modulations of the groove. The groove in long play
records has a width of 0·002″ to 0·003″ (0·05 to 0·08 mm) measured at the top of the
groove. The stylus has a radius of 0·0005″ (0·013 mm) and sits in the groove as in
figure 10.16. When the frequency is high the stylus does not follow the groove walls,
as can be seen in figure 10.17(a), and it can be shown that distortion will occur. To
reduce this a bi-radial or elliptical stylus can be used, having a cross-section as shown

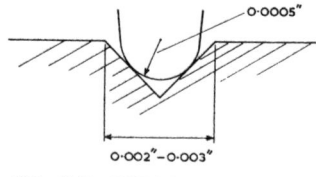

FIG. 10.16. STYLUS AND GROOVE

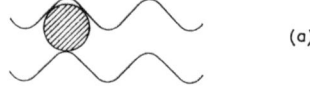

(a)

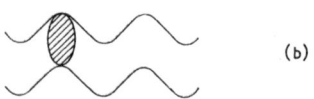

(b)

FIG. 10.17. SPHERICAL AND ELLIPTICAL STYLI

at (b), when it will be seen this tends to follow the groove better. The minor axis radius is 0·0002″ to 0·0004″ (0·005 to 0·01 mm) and the major axis radius 0·0007″ (0·018 mm). This type of stylus is more difficult to make so it is more expensive. Another type of stylus, called a Shibata, is sometimes used. This has sides which are straighter (see figure 10.18), and therefore the area of contact with the walls is increased and the pressure reduced, so that record wear is reduced.

The stylus may be made of sapphire or diamond. The better pick-ups use a diamond, as the life is longer than that of a sapphire.

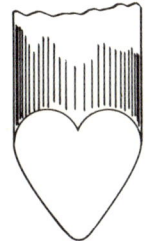

FIG. 10.18. SHIBATA STYLUS

TRACKING

Attention will now be paid to the pick-up arm. When a record is cut initially it is done on a lathe with the cutter moving across the disc in a radial path, being driven by a lead screw. For correct reproduction the pick-up should follow the same path, *i.e.* the pick-up should move across the disc following a radial path. This is difficult to do, and normally the pick-up is attached to an arm and the path is then curved as in figure 10.19. Obviously, there is quite an error which can be reduced by a long arm, but more commonly by setting the pick-up at an angle to the arm (known as head offset) as in figure 10.20. At the same time the arm is of such a length that the pick-up falls on the far side of centre (known as overhang). It will now be seen that pick-up is approximately at right angles to a radial line (and hence in line with the groove) as

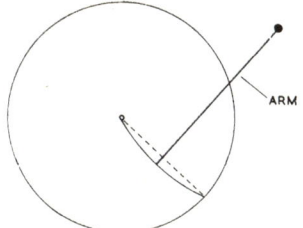

FIG. 10.19. TRACKING ERROR PRODUCED BY STRAIGHT PICK-UP ARM

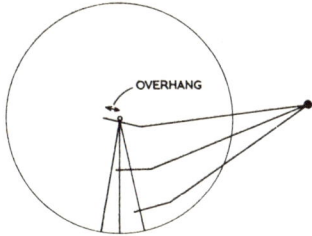

FIG. 10.20. REDUCTION OF TRACKING ERROR BY HEAD OFFSET AND OVERHANG

required; a considerable improvement on figure 10.19, where the pick-up is not in line with the groove for much of the time. There is still some tracking error, but this can be corrected by suitable design.

A parallel tracking design is now available (B & O) where the pick-up is attached to a thin ($\frac{1}{4}''$ square) short arm which is driven by a lead screw. The basic arrangement is shown in figure 10.21. The problem is to move the arm along the record at the correct speed. This is complicated by the fact that the grooves per cm are not the same

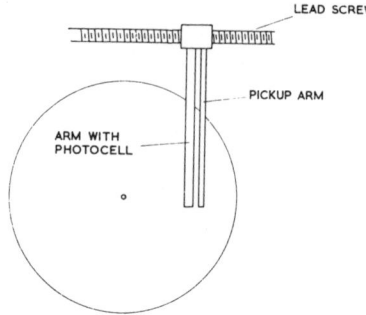

FIG. 10.21. PARALLEL TRACKING MECHANISM (B & O)

for all records and may vary on a single record. Hence some servo control system has to be used that senses when the pick-up lead screw is not driving the arm at the correct speed. The arm has a few degrees of free movement, movement from the central position is detected by a lamp and photoelectric cells. Thus the lead screw is driven by a small motor so that the arm is almost in its central position at all times, hence the tracking is radial. The whole turntable is very sophisticated so that damage cannot be done. There is another arm with a photoelectric cell next to the pick-up arm which detects the edge of the record and prevents the pick-up being lowered if there is no record. If there is no record the light is fluctuating because of the support rubbers on the turntable, and these fluctuations are detected to prevent lowering of the pick-up. The turntable operates automatically (one record) and will play 12″, 10″ and 7″, the speed automatically changing from 33 to 45 r.p.m. if it detects a 7″ record. Manual override is available and the performance is outstanding but obviously the price is high.

SIDE-THRUST

The effect of the offset of the pick-up is to produce a force tending to move the pick-up to the centre. This is illustrated in figure 10.22. The friction force between the stylus and the record now operates as shown, and since this force does not go through the arm bearing it results in a torque on the arm, tending to move it inwards. If this is not corrected for, it means that there is a steady force on one wall of the groove and the stylus possibly moved from its correct central position. Various devices are

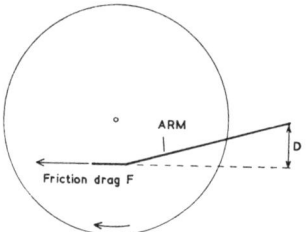

FIG. 10.22. SIDE-THRUST ON PICK-UP

used to compensate for this, such as a small weight attached to the pick-up arm by a thread passing over a small fixed rod. These devices are sometimes called "anti-skating" devices. This device should be set according to the manufacturer's instructions and its setting depends on the tracking force used. Obviously, the bearing friction of the arm must be extremely low or a side-thrust will also be produced by the arm bearing.

TURNTABLE DRIVE

The turntable is normally driven by a small induction motor (or d.c. motor in battery-operated equipment) and such motors are described in Chapter 11. The drive may be by a belt between motor and turntable or a rubber disc operating on the inside edge of the turntable. It is not possible to deal with the various arrangements in this book, but commonly at least two speeds can be selected 45 and 33 r.p.m. and in other cases three speeds 78, 45 and 33 r.p.m. Sometimes limited speed variation is available. Some turntables are manual, others play a single record automatically (lowering the pick-up, lifting it at the end and returning it to the rest position, and switching off the motor). There are also automatic record players which will play a number of records without attention. The need for these has diminished with the introduction of long play (33 r.p.m.) records, which play for 15–20 minutes a side. Some turntables use electronic drive, the turntable being driven from a motor fed from an oscillator; some turntables make use of a special low speed motor so that direct drive is possible.

The turntable should be reasonably heavy and preferably non-magnetic, with good bearings so that it runs accurately, the weight tending to reduce speed variations. The main properties quoted are "wow and flutter": "wow" is slow variations in speed and "flutter" rapid variations in speed. The other property is rumble, usually from the motor, and is the voltage output from the pick-up with no groove modulation.

PREAMPLIFIERS

If a ceramic pick-up is used then its output is proportional to the amplitude of the groove modulations. As already explained, the recording is approximately of constant amplitude, and no correction is required. That required for the portion when constant velocity recording is used can be built into the pick-up design. However, the output impedance is high, hence a preamplifier is required with an input impedance of 1 to 2 megohms. The input impedance of a normal common emitter transistor amplifier is around 3000 ohms, and therefore not suitable for direct connection. There are three solutions:

(1) Use of emitter-follower.

(2) Use of negative feedback to increase input impedance.

(3) Use of potential divider circuit.

(1) A typical circuit is given in figure 10.23. R_1 provides bias and must have a high value, say, 2 to 3 MΩ. R_2 is the emitter load resistor, say 10–15 kΩ. C_1 blocks the d.c. off from the pick-up. The input impedance of an emitter-follower is high (approximately $h_{fe} \times$ emitter load resistor, where h_{fe} is current gain) and so the load on the pick-up can be made of suitable value, commonly not less than 1 MΩ.

(2) This is similar to (1), which is an extreme case of (2). An example is shown in figure 10.24. In this circuit bias is provided by R_1 (and also some negative feedback from the collector); C_1 is a d.c. blocking capacitor. The collector load is R_2 and n.f.b. is provided by the emitter resistor R_3, which will increase the input impedance (particularly if a transistor with a high current gain is used) but also will reduce the gain.

(3) The output from a crystal or ceramic pick-up is large, say, several volts for crystal and up a volt for ceramic, and it is therefore possible to use a potential

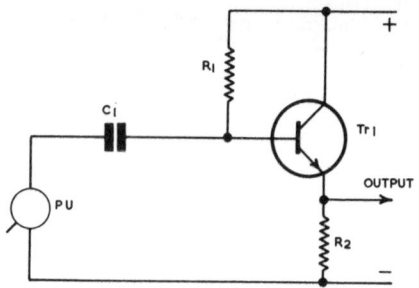

FIG. 10.23.   PREAMPLIFIER USING EMITTER-FOLLOWER

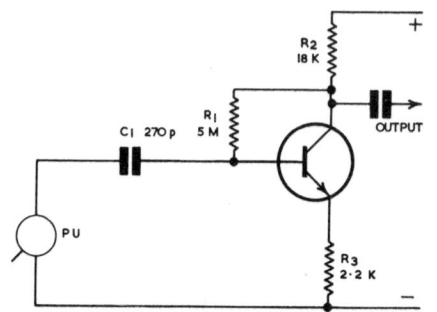

FIG. 10.24.   PREAMPLIFIER USING NEGATIVE FEEDBACK

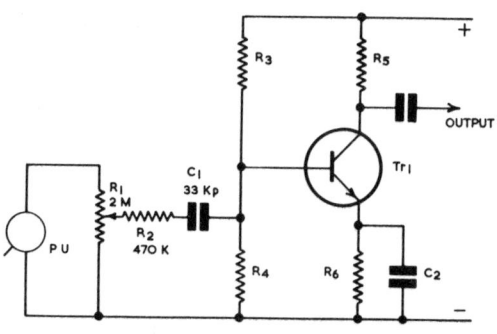

FIG. 10.25.   PREAMPLIFIER USING POTENTIAL DIVIDER

divider to prevent the amplifier loading the pick-up excessively. One circuit is shown in figure 10.25. Across the pick-up is the high resistance volume control R_1. R_2 and R_4 (together with R_3 and the input impedance of Tr_1) form a potential divider which will reduce the voltage, but the minimum load on the pick-up is approximately R_1 and R_2 in parallel. This is perhaps lower than desirable, but will occur only when R_1 is at the maximum volume position. Placing the volume control before the amplifier prevents overloading. In some cases the tone controls are placed across the pick-up, and in some stereo equipment also the balance control. The complete amplifier may be very simple; as the first stage may be sufficient to operate the driver stage of the main amplifier, four transistors may be sufficient. An integrated circuit may be used, again using a potential divider to feed it, and a single integrated circuit is often used acting as a preamplifier and output stage.

Instead of using an amplifier with a high input impedance a value of 47 kΩ (the same as for magnetic) may be used. The level response is now upset, the high frequency response increased, and the response becomes similar to that of a magnetic pick-up and the same preamplifiers can be used.

In a magnetic pick-up the output is proportional to velocity as shown by the RIAA curve of figure 10.3. In order to correct for this the amplifier must have a frequency response that is the inverse of this and is shown in figure 10.26, thus the amplifier must have its greatest gain at low frequencies. The change in gain is

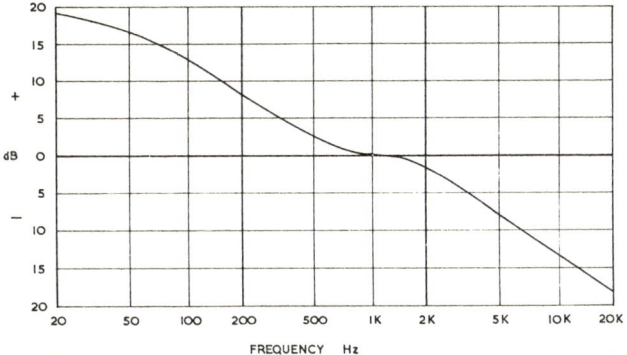

FIG. 10.26. CHARACTERISTIC OF EQUALIZING AMPLIFIER FOR MAGNETIC PICK-UP

large and most commonly obtained by the use of a frequency selective negative feedback circuit, one circuit being given in figure 10.27. Magnetic pick-ups are designed to operate into a load of 47 kΩ, hence R_1 together with the input impedance of the

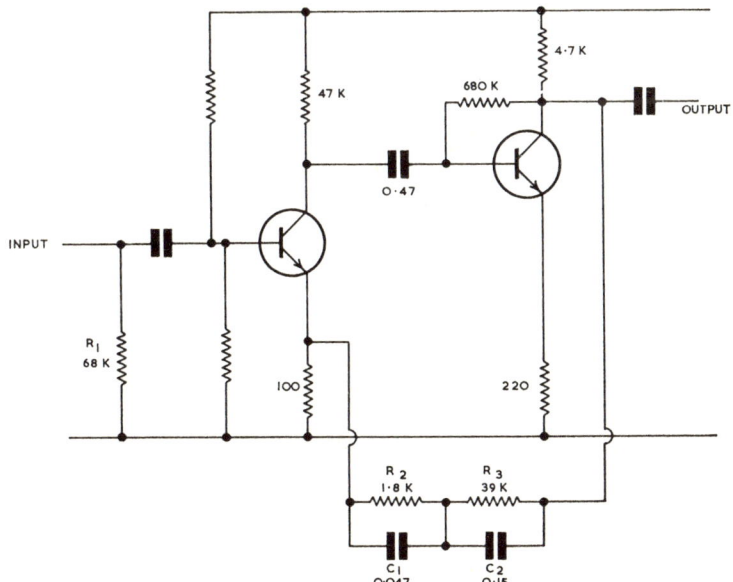

FIG. 10.27. TYPICAL CIRCUIT OF EQUALIZING AMPLIFIER FOR MAGNETIC PICK-UP

amplifier provide this load. The amplifier is fairly standard apart from the feedback circuit $R_2 R_3$ and $C_1 C_2$ and these should be small tolerance components. At low frequencies the reactance of C_1 will be high (at 100 Hz it is about 30 kΩ), and feedback is through $R_2 R_3$ and C_2. At low frequencies the reactance of C_2 will be high (at 100 Hz about 10 kΩ), and the feedback is small and the gain large. (R_3 prevents the feedback becoming very low at very low frequencies and giving excessive gain). As the frequency is increased the reactance of C_2 decreases, so increasing the feedback and reducing the gain. This gives the lower frequency part of the characteristic. As the frequency increases, the reactance of C_1 becomes comparable (at 1 kHz it is about 3 kΩ), and then low, compared with R_2 and so increasing the feedback and reducing the gain (at 10 kHz the reactance of C_1 is 300 ohms). In some preamplifiers d.c. feedback is used and a d.c. coupled amplifier but the principle is not altered.

Care is necessary in the design of this amplifier or distortion can occur because the input is large at high frequencies, while at middle frequencies the gain must be sufficient to feed the main amplifier (or preamplifier and tone control stage). The overall change of gain from low to high frequencies is about 40 dB (100/1) and distortion may also occur at low frequencies where the feedback is small. The amplifier should be able to accept an input of at least 100 mV at 1 kHz without distortion, and, of course, corresponding voltage levels at other frequencies.

CHAPTER 11

TAPE RECORDING AND RECORDERS

THERE are two methods of storing music and speech: on the record or disc (considered in Chapter 10) and the magnetic tape. The latter has the advantage that recording is easy and that the tape can be re-used. In this chapter we shall deal first with the principles of tape recording and later with recorders.

The basic idea of tape recording is simple, that of magnetizing a wire or tape and then using the remaining magnetism to induce an e.m.f. in a coil so that it can be played back. Although the idea is simple many problems had to be overcome before such a system was of good enough quality for general use. Originally, iron wire or tape was used, which is now replaced by plastic tape with a magnetic coating (common iron oxide) but now other materials are being used.

In order to magnetize the tape a head, as it is called, is used and is constructed as shown in figure 11.1. The ring is made of magnetic material such as mumetal, permalloy or ferrite, and has a magnetic air-gap. The gap is not a physical one as it

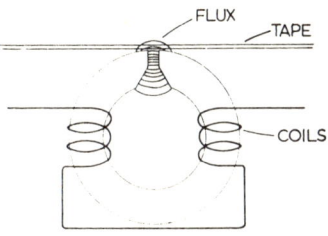

FIG. 11.1. TAPE HEAD

would soon fill with dirt and oxide. Hence, the gap is produced by inserting a piece of non-magnetic material such as gold, phosphor bronze or glass. At the air-gap the magnetic field produced by the coil fringes out and it is this fringing field that magnetizes the tape. In practice the tape is moved at constant velocity past the head and the recording current in the coil causes a varying field, resulting in a varying magnetism in the tape. For an alternating current the resulting tape magnetization is as figure 11.2. If the frequency is low the result is as at (a), and at a higher frequency as at (b). Thus the higher the frequency the nearer the poles are together and the slower the

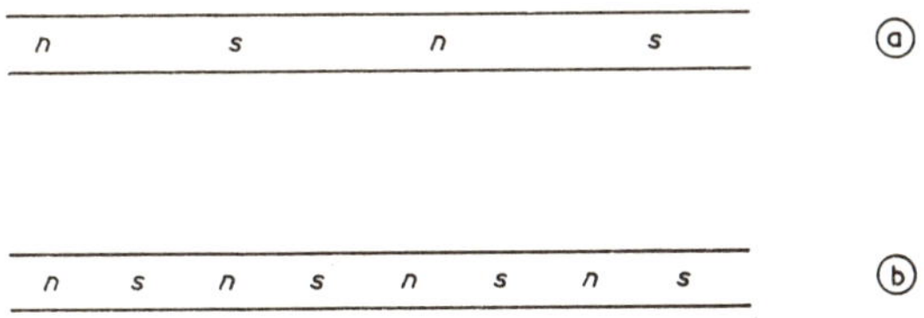

| n | s | n | s | (a) |

| n | s | n | s | n | s | n | s | (b) |

FIG. 11.2. MAGNETIZATION OF THE TAPE

tape speed the nearer the poles. If this magnetized tape is now passed over a head the varying flux will induce an e.m.f. in the coils and we get playback.

The recording process is, of course, closely connected with the properties of the tape, and to some extent of the head. Unfortunately, the magnetic characteristics are non-linear and there is also a hysteresis effect. A typical graph showing the relationship between magnetizing force H (in ampere-turns/metre) and the resulting flux density B (in Tesla or Webers/m^2) is shown in figure 11.3. If we start from a non-magnetized material at A and increase the value of H, then flux density B rises slowly at first up

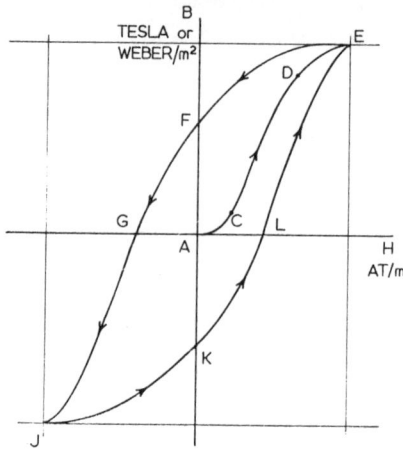

FIG. 11.3. B-H LOOP OF MAGNETIC MATERIAL

to point C, but from C to D the rise is more rapid, and beyond C saturation occurs
and the rate of rise is decreased. If now the value of H is reduced, the flux density does
not follow the same curve but lags behind the changes of H and there is hysteresis.
When H is zero the value of B is given by the point F, this being known as the rema-
nent flux density or retentivity. In order to reduce B, H must be reversed and at some
point G the value of B is zero. The value of H at this point is known as the "coercive
force". If H is increased further then B increases in the opposite direction to point J
(corresponding to point E in the opposite direction). The other half of the loop is
identical. The whole curve is known as a "hysteresis loop". The shape of the loop will
depend on the material, but also the maximum flux density (corresponding to point E).
The greater the flux density the greater the saturation and the more curved the loop.
In the recording process the two magnetic portions to be considered are the tape
head and the tape itself. The material of the head will have a hysteresis loop but can
be made more linear by the introduction of another gap at the back of the head, and
so we will neglect the non-linearity of the head.

As regards the tape, one might expect that it would be left with a magnetism corresponding to point F (or K) in figure 11.3. This would be so if the tape magnetic
material formed a closed magnetic circuit. It does, of course, have a large air-gap and
is similar to a bar magnet. When there is an air-gap the remaining magnetism has to
force flux through the air and the flux is less than with a closed circuit—in fact the
value of the flux is somewhere between points F and G, shown in figure 11.4. The tape
operates at a point such as M with a resulting flux ϕ_M (flux = flux density B × cross-
sectional area), which is less than the flux at F. Now consider a tape that has not been
previously magnetized (*i.e.* contains no remaining magnetism) but subjected to
various values of magnetizing force, as shown in figure 11.5. If the tape is magnetized
to point E then the resulting magnetism will correspond to point M, where the curve
cuts the line OZ. The position of OZ depends on the effective air-gap. If the tape is
only magnetized to point N then the remaining magnetism corresponds to point P,
and so on. Thus the resultant magnetism is given by points M, P, Q, etc. If the flux
corresponding to these points is plotted against H the result is shown in figure 11.6.
Although this may look like a normal B-H curve it is important to realise that it is
something quite different, notwithstanding that it happens to have a similar shape.
Now suppose that a sinusoidal current is passed through the coils on the head: this
results in a sinusoidal magnetizing force H as shown. From this graph we can obtain
the shape of the resulting magnetic flux waveform, when it is seen that it is far from
sinusoidal. Obviously, this will induce a distorted waveform in the playback head.

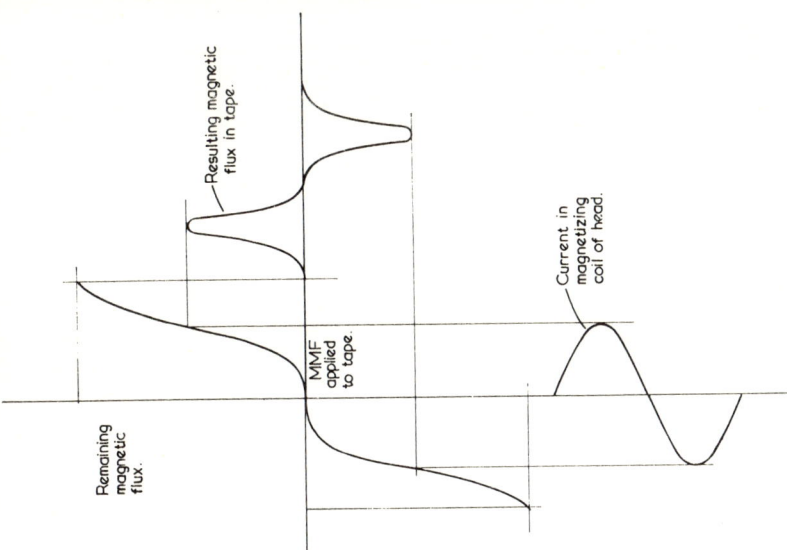

FIG. 11.6.   RELATIONSHIP BETWEEN MMF AND RESULTING FLUX

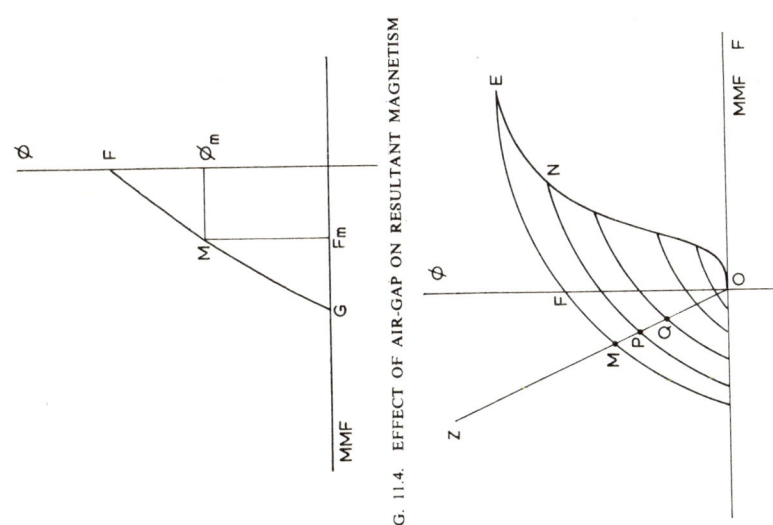

FIG. 11.4.   EFFECT OF AIR-GAP ON RESULTANT MAGNETISM

FIG. 11.5.   REMAINING MAGNETISM ON TAPE

This system is useless, the distortion is very high and the sensitivity is poor.

To overcome this basic problem a bias is used, which may be d.c. or high frequency, a.c.

D.C. BIAS

If we apply a steady magnetization to the head by, say, passing a direct current through the coils of the head, then the resulting flux is ϕ_{DC} as shown in figure 11.7. ϕ_{DC} is arranged to be about the centre of the fairly straight part of the characteristic.

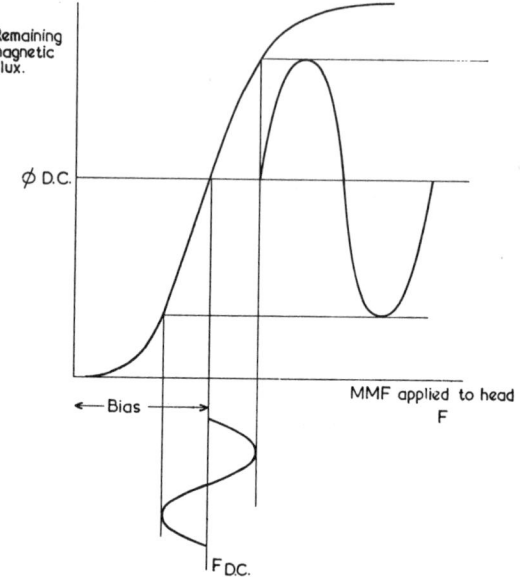

FIG. 11.7. USE OF D.C. BIAS

If now a sine wave is added to this d.c. we get a variation of flux as shown, being approximately sinusoidal. There is now little distortion and the system becomes practicable. However, it is not good as it leaves the tape with a d.c. magnetization. If this were constant it would not matter; but owing to the nature of magnetism (which consists of a number of small magnets), considerable noise is produced. Thus, the signal-to-noise ratio using this system is poor.

HIGH FREQUENCY BIAS

Instead of using a d.c. bias a high frequency bias is now used, which is the reason for the present high performance of tape recorders. There does not appear to be a really satisfactory explanation; a simple explanation is given in figure 11.8. By adding the bias to the signal the fairly straight part of the B-H characteristic is used. It is important to note that the bias and signal are simply added together; the bias is NOT modulated by the signal. The bias frequency is normally from 50 to 100 kHz and is much larger in amplitude than the signal.

For best results the magnitude of the bias current is rather critical; the way in which the characteristics change with bias current is shown in figure 11.9. As the bias is increased the output increases rapidly at first and then reaches a peak. After that it slowly falls. However, the amount it falls depends on the recording frequency, falling more rapidly for recordings of high frequency. The distortion is high at low

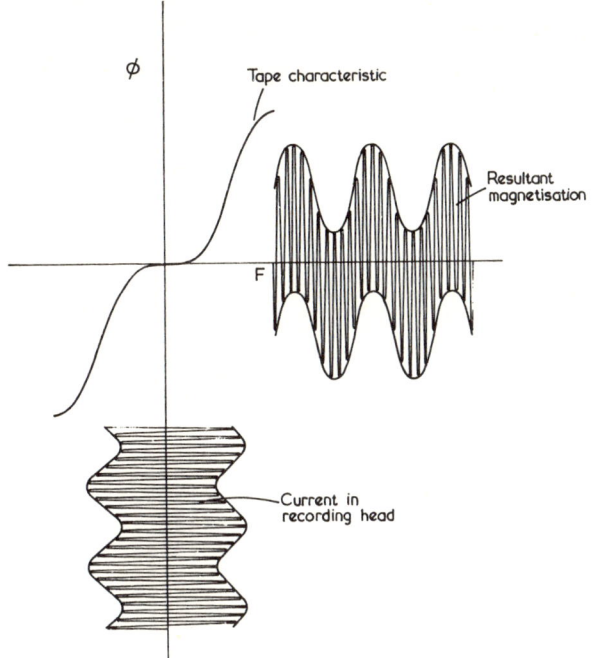

FIG. 11.8.   USE OF HIGH FREQUENCY A.C. BIAS

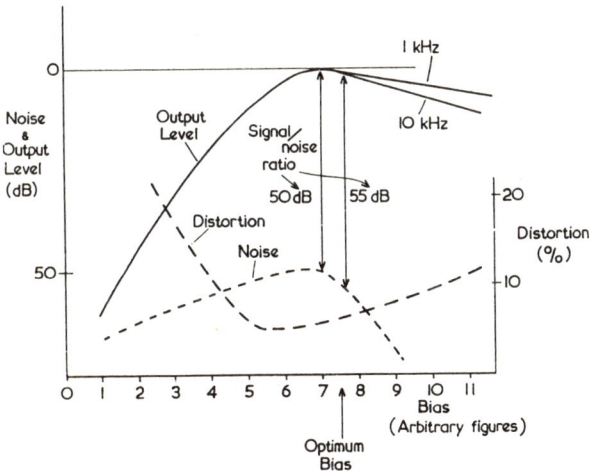

FIG. 11.9.   EFFECT OF MAGNITUDE OF HIGH FREQUENCY BIAS

bias values and reaches a minimum, which does not correspond with the bias for maximum output. The distortion rises more slowly after the minimum point. The noise at first rises and reaches a maximum at about the same bias as maximum output. For bias values above this maximum the noise drops off rapidly. The optimum bias is therefore a compromise. Usually the bias is set to give maximum output or a slightly greater bias which reduces the output by 1 to 2 dB. The latter point gives a better signal-to-noise ratio (see figure 11.9) because the noise drops off more rapidly than the signal. If the bias is increased too far the frequency response becomes poor. The waveform of the bias current is important; it must be free from distortion, particularly even harmonics, which tend to leave the tape with permanent magnetism which increases the noise.

RECORDING PROCESS

We may consider the tape as consisting of a large number of small magnets, as shown in figure 11.10. When the tape is unmagnetized these are arranged in a random way, as shown at (a). The magnets therefore cancel each other out and produce no

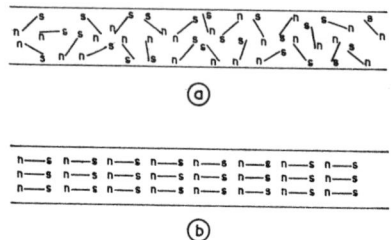

FIG. 11.10. MAGNETIZATION OF THE TAPE

external magnetism (apart from the noise referred to under d.c. bias). When the tape is magnetized and fully saturated they are all aligned as at (b). Consider now the case where the wavelength of magnetism on the tape (*i.e.* distance between like poles) is large compared with the head gap, as in figure 11.11. When the magnetism from the coils on the head is a maximum the magnets are aligned in one direction, as at A.

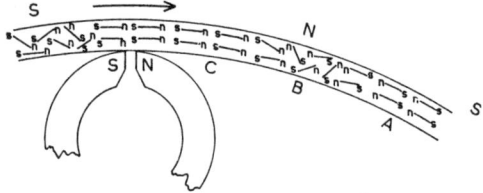

FIG. 11.11. MAGNETIZATION OF THE TAPE WHEN WAVELENGTH IS LONG COMPARED WITH AIR-GAP

When the current is zero they remain in the random fashion, as at B. When the current is reversed they are aligned in the opposite direction, as at C. Thus N and S poles are produced by the tape. Consider now the case where the wavelength is comparable to or less than the gap in the head. This is shown in figure 11.12. As a portion of the tape passes over the gap the head magnetization may reverse several times, but what

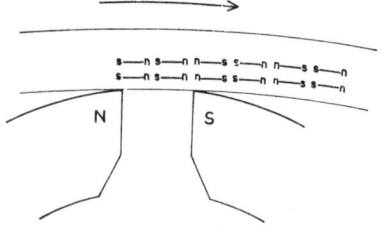

FIG. 11.12. MAGNETIZATION OF THE TAPE WHEN WAVELENGTH COMPARABLE WITH THE AIR-GAP

determines the final magnetization of the tape is the polarity of the trailing pole tip (RH one) at the instant the tape leaves it. What happens during the period when crossing the gap has little effect. The process is obviously very involved. Because of the above the width of the recording head is not too important.

REPLAY

Once the tape has been recorded it is passed over a replay head, as shown in figure 11.13. The reluctance of the head is less than the air, therefore some flux passes through the head as shown. As the tape passes over the head this flux will vary in magnitude

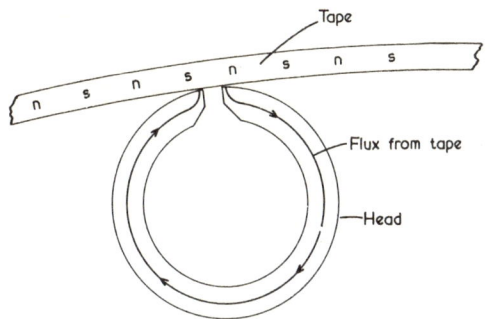

FIG. 11.13. PLAYBACK HEAD

and direction. Thus, if a coil or coils are placed on the head the flux will cut the coils and induce an e.m.f. in them. This e.m.f. will be proportional to the **rate of change of flux**, and not the actual value of flux. Since the rate of change of flux will be proportional to frequency (other factors being constant) the e.m.f. will be proportional to frequency. However, two factors upset this relationship. Suppose that the wavelength on the tape is equal to the gap width, as shown in figure 11.14. In this case it is seen that two similar poles are opposite the poles of the head, and no flux flows in the head. Therefore, at a

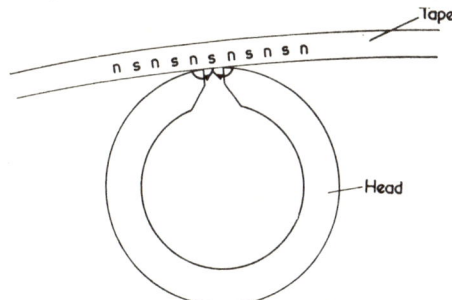

FIG. 11.14. PLAYBACK HEAD WHEN THE WAVELENGTH IS EQUAL TO GAP LENGTH

frequency corresponding to this wavelength, there is no e.m.f. produced in the coil. This is known as the "extinction frequency". The relationship between frequency and output is shown in figure 11.15. Over part of the range the output is proportional to frequency, but then drops off rapidly to zero at the extinction frequency. It rises again above this frequency, but this fact is of no value. If we now consider the case where the wavelength is large compared with the gap width the effect is as in figure 11.16. It will be seen that now some of the flux does not pass through the coils and the e.m.f. is less than it should be. Thus, at low frequencies, the output drops off more rapidly, as shown in figure 11.15. Obviously some compensation is necessary in the playback amplifier to correct for this characteristic. The gap width of the replay head (which may be the

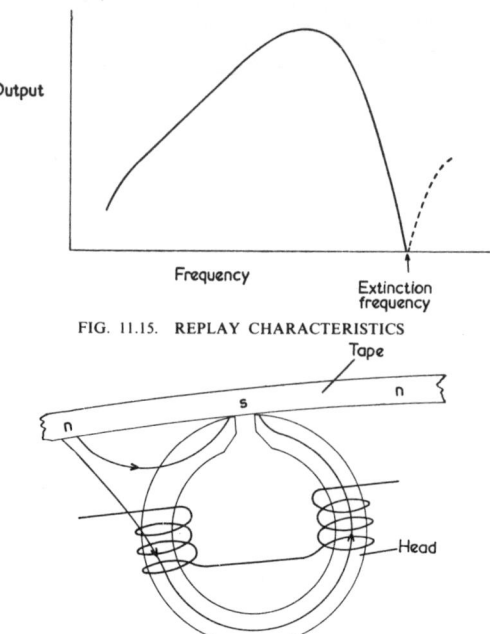

FIG. 11.15.   REPLAY CHARACTERISTICS

FIG. 11.16.   WAVELENGTH LONG COMPARED WITH THE GAP LENGTH

same head as the record head) is now vitally important, and, if good high frequency response is required, should be made extremely small.

WAVELENGTH ON TAPE

It is useful to see the order of magnitude of the wavelength of the magnetization on the tape. Consider a tape speed of 19 cm/second ($7\frac{1}{2}''$/second) and a frequency of 1 kHz. In the time of one cycle the tape will have moved 19/1000 cm = 0·019 cm or 0·19 mm (approximately 8/1000 of an inch). When the frequency is raised to 10 kHz this is reduced by a factor of 10 and becomes 0·019 mm (0·8/1000 of an inch) which is very small. As the tape speed is reduced these figures are also reduced, the wavelength being proportional to the tape speed.

ERASING

One of the great virtues of tape recording is that any recorded material can be erased from the tape and the tape used again. The old material can be erased by the application of a strong d.c. field which takes the tape into saturation. However, this leaves the tape with a steady d.c. magnetization and results in a poor signal-to-noise ratio. It is only used in a few cheap recorders, and a permanent magnet may be used.

Erasing is normally done by a high frequency field, the same one as used for bias. The erase head is placed before the recording head so that any previous recording is automatically erased. For complete erasure the tape should be taken through several cycles of magnetization up to saturation and then the magnetization gradually reduced to zero. This can be done by using a head with a relatively large gap. As the tape enters the field at A (figure 11.17) the field builds up and reaches a maximum at the centre of the gap corresponding to B. The field must be strong enough at this point to saturate the tape. The field is then gradually reduced to zero as it leaves the gap. Figure 11.18 shows what is happening as regards the B-H loop. From A to B (figure 11.17) the flux is increasing, hence the tape is taken through larger and larger B-H loops, as shown in

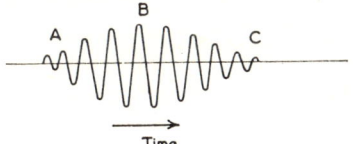

FIG. 11.17. ERASURE BY HIGH FREQUENCY FIELD

figure 11.18(a). At B it is taken round a large loop as at (b) and then from B to C it is gradually reduced to zero as at (c). The erase current should be of good waveform and not have even harmonic content, as this tends to leave the tape with some magnetism which produces a poor signal-to-noise ratio. In order to fully saturate the tape quite large erase currents are required and, to ensure that the head does not saturate it is commonly made of silicon steel rather than mumetal. A ferrite core may also be used. In some cases two gaps are used so that two attempts are made at erasing. Instead of erasing the tape on the tape recorder it may be bulk erased, as it is called. In this case the whole spool of tape is erased at once. This is done on a bulk eraser, which consists of a coil connected to the 50 Hz mains having a laminated open iron core so that there is a large external field. The spool of tape is placed near the coil and moved around, and then taken slowly several feet away before switching off the coil. Bulk erasing often results in a better signal-to-noise ratio than erasing on the tape recorder.

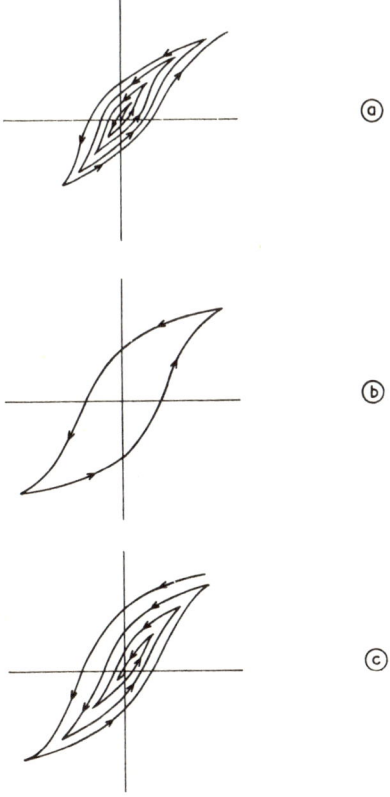

FIG. 11.18. ERASING PROCESS

BASIC TAPE RECORDER

Two basic electrical arrangements are used, which are shown in figure 11.19. That shown at (a) is known as the three-head recorder. On record the erase head is fed from the high frequency oscillator, which also supplies bias to the record head. The signal

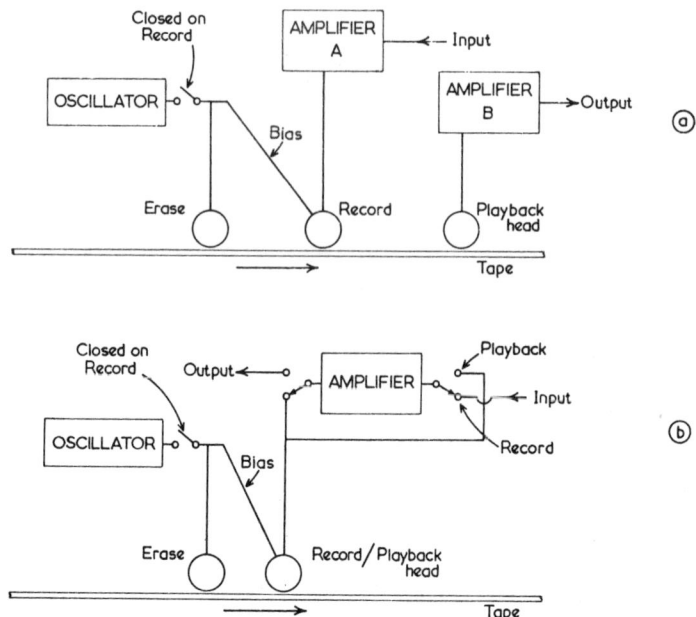

FIG. 11.19. BASIC ARRANGEMENTS OF RECORDER

input is fed to amplifier A which feeds the record head. The playback head is connected to amplifier B and gives the output. The advantage of this arrangement is that the recording can be monitored by the playback head as it is recorded. However, two amplifiers are required, which makes it more expensive. Another advantage is that the record head and playback head can be designed for their specific purpose rather than a combined purpose. At (b) is shown the arrangement of a two-head recorder. As before, on record the oscillator feeds the erase head and bias for the record head. The input is fed into the amplifier which feeds the record head. For playback both switches are operated so that the record/playback head feeds the amplifier which produces the signal output. Obviously the recording process cannot be monitored, but only one amplifier is required. It is necessary to change the characteristics of the amplifier between record and playback. This is the most common arrangement in domestic tape recorders but three-head recorders are readily available.

The basic mechanical arrangement is given in figure 11.20. The tape comes off a supply spool and over the heads, then pulled across the heads by a motor-driven capstan, *i.e.* a polished steel cylinder. So that the capstan drives the tape a rubber pressure roller is used to press the tape against the metal capstan. The tape is then fed to the take-up reel. The take-up spool must be driven and the supply reel must keep the tape taut over the heads. The speed of the tape is settled by the capstan diameter and speed. More details will be given later.

In order to locate a part of the tape, arrangements are made to drive the tape rapidly forwards or backwards by lifting the pressure roller off the capstan, the tape then being driven by the appropriate spool or reel.

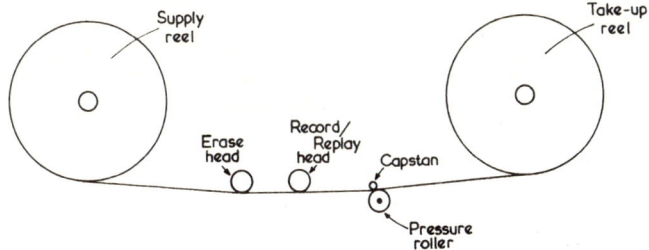

FIG. 11.20. BASIC MECHANICAL ARRANGEMENTS OF RECORDER

TAPE SPEED AND TRACKS

The greater the tape speed the higher the extinction frequency because the wavelength on the tape is proportional to tape speed. Thus the higher the tape speed the higher the frequency response. However, the faster the tape speed the more tape is used for a given recording. Thus a number of tape speeds are used depending on the purpose.

The normal tape speeds are:

Tape speed		Approx. max. frequency
cm/second	inches/second	
38	15	over 20,000 Hz
19	$7\frac{1}{2}$	20,000 Hz
9·5	$3\frac{3}{4}$	18,000 Hz
4·75	$1\frac{7}{8}$	11,000 Hz
2·4	$\frac{15}{16}$	5,000 Hz

Approximately maximum frequency response is also given, but obviously depending on the quality of the recorder and particularly on the gap width of the playback head. The figures quoted are for good recorders and cheap ones may be considerably below these figures. A speed of 38 cm/second (15 inches/second) is mainly used for professsional recording. The speed of 19 cm/second ($7\frac{1}{2}$ inches/second) is used on domestic recorders where good frequency response is required, but 9·5 cm/second ($3\frac{3}{4}$ inches/second) is commonly used and gives satisfactory recording of speech and music. It is also the speed used in cartridges (see later). 4·75 cm/second ($1\frac{7}{8}$ inches/second) is used where tape economy is important. It gives satisfactory results on speech, but a rather limited frequency response for music. This speed is also used in cassettes (see later). 2·4 cm/second ($\frac{15}{16}$ inch/second) is only satisfactory for speech recording, but, of course, is very economical in tape usage.

The material used for tape recording is a plastic tape with a suitable magnetic coating. The thickness of the plastic tape varies from 0·052 mm (0·002″) to 0·018 mm (0·0007″) and the coating is 0·013 mm (0·0005″) to 0·009 mm (0·0004″). The normal tape used in reel-to-reel machines is $\frac{1}{4}$″ wide. Wider tapes are used professionally on multitrack recorders and video recorders. The general purpose tape is PVC (polyvinyl chloride) with a thickness of 0·052 mm (0·002″). Thinner tapes are used, the first being known as long play. This is 0·035 mm thick and, for a given spool size, gives 50% greater playing time because a greater length of tape can be wound on the spool. Double-play tape is thinner still (0·026 mm) and this allows twice as much tape to be wound on a given spool size. There is also triple-play tape (0·018 mm) which, of course, gives three times the playing time for a given spool size. This is very thin and not commonly used. The thinner tapes are normally made of polyester rather than PVC. Tapes thinner than triple play are sometimes used in cassettes.

On domestic equipment spool sizes vary from 8 cm (3″) to 18 cm (7″), but not all tape recorders will take the 18 cm size, many only going up to 13 cm (5″). The spools are similar to those used for standard 8 cine film. Super 8 cine spools are similar, but with a larger hole in the centre. Professionals use larger spools, the European one

TABLE 11.1

LENGTH OF TAPE

TYPE OF TAPE	SPOOL DIAMETER															
	8 cm 3 inch		10 cm 4 inch		13 cm 5 inch		15 cm 5¾ inch		18 cm 7 inch	
	m	ft	m	ft	m	ft	m	ft	m	ft
Normal	45	150	90	300	180	600	270	900	360	1200
Long play	65	210	135	450	270	900	360	1200	540	1800
Double play	90	300	180	600	360	1200	540	1800	730	2400
Triple play	135	450	270	900	540	1800	730	2400	1080	3600

being 30 cm (11½″) diamter and has a single flange. NAB spools of 26·5 cm (10½″) and 36 cm (14″) are used, some of which having detachable flanges. Details of spool sizes and lengths of tape on the spools are given in Table 11.1. In Table 11.2 are given the playing times for various lengths of tape.

TABLE 11.2

PLAYING TIME PER TRACK (MINUTES)

TAPE LENGTH m	ft	19 cm/s 7½ inches/s	9·5 cm/s 3¾ inches/s	4·75 cm/s 1⅞ inches/s	2·4 cm/s 1⅝ inch/s
45	150	3·8	7·5	15	30
65	210	5·5	11	22	45
135	450	11	22	45	90
180	600	15	30	60	120
270	900	22	45	90	180
360	1200	30	60	120	240
540	1800	45	90	180	360
730	2400	60	120	240	480

TO GET TOTAL TIME:

Full track (single)	× 1
Two track mono	× 2
Two track stereo	× 1
Four track mono	× 4
Four track stereo	× 2

The usual tape coating is iron oxide (ferric oxide), but other materials are now available, particularly for use on cassettes. One of these is chromium dioxide, but this requires a different bias setting and/or frequency compensation to iron oxide. Other proprietary mixtures are available. One consists of two coatings, iron oxide with a top coating of chromium oxide. This does not require a change in bias or compensation.

Turning now to the tracks, professional recordings usually use the whole width of the tape and is known as single track. This is shown in figure 11.21(a). The larger the amount of tape used the greater the voltage induced in the head and the better the the signal-to-noise ratio. All tapes have some defects and areas where the oxide coating is less magnetic. This results in the signal dropping in level and is known as "drop outs". The greater the width of tape used the fewer the drop outs. For many

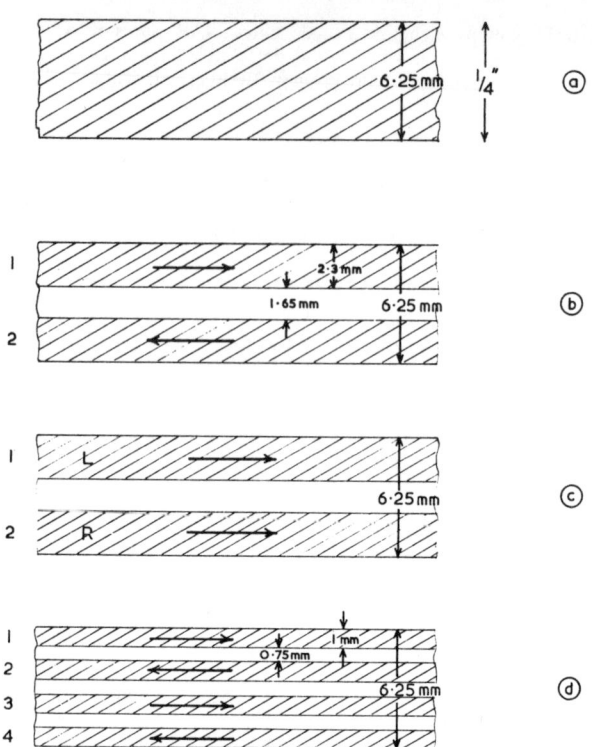

FIG. 11.21. TRACKS OF TAPE. (a) SINGLE; (b) TWO TRACK MONO; (c) TWO TRACK STEREO; and (d) FOUR TRACK

purposes the whole width of tape need not be used and two or twin tracks are used, as shown at (b). For recording in mono a recording is made on track 1, then the tape is turned over and a recording made on track 2 (which is now at the top) in the opposite direction. The same head is used for both recordings. When a stereo recording is required the two tracks are recorded at the same time as at (c), the upper track being the left-hand (L) signal and the lower track the right-hand (R) signal. In this case a two-track head is used and the tape run only in one direction. For domestic use four-track recording is now the most common, shown at (d). The arrangement of the head is shown in figure 11.22, there being heads opposite tracks 1 and 3. For mono recording track 1 is first recorded using the upper head. The tape is then turned over and track 4 is recorded in the opposite direction. The tape is now run in the original direction but using the lower head, so that track 3 is recorded. The tape is then turned over and track 2 is recorded in the opposite direction using the same head. A switch must be fitted so the record amplifier can feed the upper or lower head. For a stereo recording tracks 1 and 3 are recorded together (using both heads). The tape is then turned over and tracks 2 and 4 are recorded in the opposite direction. Obviously a four-track recorder

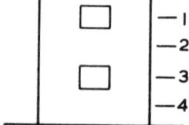

FIG. 11.22. HEADS ON FOUR-TRACK RECORDER

can record twice as much material as a two track for a given length of tape. However, the signal-to-noise ratio is not as good and there are likely to be more drop outs.

Two-track and four-track recorders are not compatible and great care has to be taken when using a mixture of two-track and four-track recorders. In particular a four-track recording in stereo cannot be played in mono on a two-track recorder, since the two stereo tracks are not adjacent. A two-track recording can be played on a four-track machine with some loss of output, but a four-track recorded tape cannot be played on a two-track machine. If only tracks 1 and 4 (or 2 and 3) have been recorded and the other tracks are clear then the four-track recording can be played back on a two-track machine. Problems can also occur over erasing.

With the introduction of quadrophonic recording (see Chapter 12) 4 tracks are required. This can be done using the 4 tracks of a four-track recording, but, of course, four heads are required. There can now be some confusion between the normal four-track recorder and the recorder capable of recording all four-tracks at the same time. To distinguish them the normal four-track recorder is called a four-track 2 channel recorder, and the other a four-track 4 channel recorder.

TAPE EQUALIZATION

Suppose we had a perfect recording system such that the remaining magnetism in the tape was proportional to the current in the head. When this tape was played back (with a perfect playback head) the e.m.f. induced in the playback head would be proportional to the rate of change of flux, and hence proportional to frequency. Hence, even with this perfect system some equalization would be required in the playback amplifier if the recording system were to have a uniform frequency response. Unfortunately this perfect recorder and playback system does not exist and there are many other effects which upset the frequency response and must therefore be corrected for. The system is complex and there is considerable difference between professional recorders and domestic recorders. Professional recorders run at relatively high speeds (38 cm/second or 15″/second) and therefore it is not difficult to obtain a response up to, say, 20 kHz. However, it is important that the recording is made to some standard so that it can be played back satisfactorily on other recorders. It must also be remembered that the standards set by professional recording engineers are much higher than those possible or required by the average domestic user. In domestic recorders the speed is kept down to reduce the consumption of tape and hence, other things being equal, the maximum frequency response is reduced. To improve the frequency response more equalization is used, and attempts are made to squeeze as much as possible out of the tape.

Consider first the problem of recording and what equalization is required. Unfortunately the remaining magnetism, for a constant recording current, is not independent of frequency. There are a number of factors, *e.g.* self-demagnetisation of the tape, penetration loss, spacing loss and head loss, which cause the remaining magnetism to be reduced at high frequencies. We have seen that on the replay side the output increases proportional to frequency over a limited range, but, at high frequencies, it drops off rapidly to zero at the extinction frequency. In order to reduce the compensation required at high frequencies in the replay amplifier these frequencies are boosted on recording and often the very low frequencies are also boosted because the replay head output is so low at these frequencies. The amount of boost must not be too great or overloading and distortion will occur.

If recordings are to be interchangeable then all should be made to some standard. Unfortunately, there are a number of standards that differ slightly. Some are, NAB (American), DIN (German), IEC (European) and British Standards (BSI). The characteristics are quoted in different and often confusing ways. For simplicity the British Standard (BSI 1 568, Nov 1971) will be used, which is similar to the CCIR standard.

The confusion that arises is due to the fact that the recording characteristics are quoted in terms of a time-constant. This is because the recording characteristics are similar in shape to the impedance-frequency characteristics of a C-R circuit. However, this time-constant has nothing directly to do with any time-constant in the compensating circuit of the amplifier. The B.S.I. states: "... that the tape should be recorded in such a way that, with a constant sinewave input to the recorder, the short-circuit magnetic flux-frequency characteristic shall have a certain shape". The term: "short circuit magnetic flux" is that flux which would flow through an ideal head with zero reluctance and assuming that there is perfect contact between head and tape. The shape of the characteristic is quoted in terms of the impedance-frequency characteristics of a C-R circuit, this circuit having a certain time-constant, *i.e.* CR product. It is not always made clear whether it is a series or parallel circuit and both are used. Consider first a parallel circuit, as shown in figure 11.23. At low frequencies the reactance of C will be high compared with R, hence the impedance is R as shown at (b). As

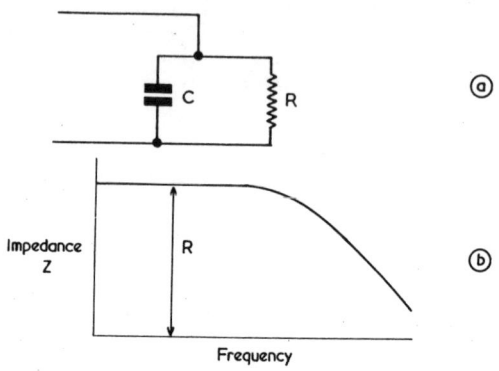

FIG. 11.23.   PARALLEL C-R CIRCUIT

the frequency is increased the reactance of C decreases, hence the total impedance decreases following a curve as shown. The shape of the curve does not depend on the values of C and R, but the frequency at which the impedance starts to fall depends on the CR product. If a series circuit as in figure 11.24, is considered, then at high frequencies the reactance of C will be low, and hence the impedance will be equal to the value of R, as shown at (b). As the frequency is decreased the reactance of C increases, and when it becomes comparable with R the total impedance rises as shown. Again,

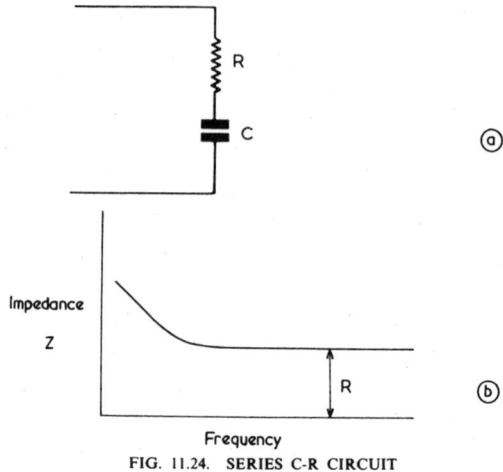

FIG. 11.24.   SERIES C-R CIRCUIT

the shape of the curve is independent of the values of C and R, but the frequency at which the impedance rises depends on the product CR (the time-constant).

When quoting the recording characteristic it is the characteristics of the parallel circuit that apply for high frequencies and the series circuit for low frequencies. The time-constant for high frequencies is t_1 and that for low frequencies is t_2. The recording characteristic will depend on the tape speed, and some figures are given in Table 11.3.

TABLE 11.3

SPEED		t_1	3 dB TURNOVER POINT	t_2	3 dB TURNOVER POINT
76 cm/s	30″/s ⎫	35 μs	4·5 kHz	Infinite	0
38 cm/s	15″/s ⎭				
19 cm/s	7½″/s	70 μs	2·5 kHz	Infinite	0
9·5 cm/s	3¾″/s	90 μs	1·8 kHz	3180	50 Hz
4·75 cm/s	1⅞″/s	120 μs	1·35 kHz	1590	100 Hz

Where the time-constant is quoted as infinite it means that the characteristic is a horizontal straight line, there being no compensation. These characteristics are plotted in figure 11.25, the characteristics being plotted in terms of dB for the value of the remaining magnetism. The proposed time-constants for the new chromium dioxide tape are 70 and 3180 μs at 4·75 cm/s (1⅞ inches/second) (IEC).

Considering first the high-frequency parts of these characteristics it is seen that there is a drop in the remaining magnetism at high frequencies and perhaps one would expect a boost from what was said earlier. In fact, these curves do represent a boost because with no compensation in the recording amplifier the remaining magnetism would drop off, say, as curve A (for the various reasons already given). The boost given by the amplifier is the difference between curve A and the standard characteristic for the same speed. One might wonder why greater compensation is not made. If it were then the tape would be overloaded and distortion would occur. At the low-frequency end there is a boost for low tape-speeds. This helps to overcome the drop that occurs at very low frequencies (due to all the flux not linking with the replay coils) and also reduces any hum. Typical amplifier characteristics are shown in figure 11.26, where it will be seen that there is quite a large increase in gain at high frequencies. The exact shape of these will depend to some extent on the record head design. If the type of tape is changed then ideally these curves should be changed, but this is normally only done for chromium dioxide tape.

As regards the playback head, under ideal conditions this produces an e.m.f. proportional to frequency, the output rising 6 dB per octave. As already shown, there is in practice a loss at high frequencies due to the finite gap width. There is also some loss at very low frequencies. Thus the amplifier characteristics must be such that the gain decreases proportional to frequency over most of the frequency range, but must rise rapidly at high frequencies. Typical characteristics are given in figure 11.27, but the exact shape will much depend on the design of the replay head. When a tape recorder has facilities for operating at a number of speeds then the characteristics of the replay and record amplifier must be changed when the speed is changed.

RECORDING AMPLIFIERS

The recording amplifier will be considered separately from the replay amplifier, for simplicity, but it must be remembered that in two-head recorders the same amplifier is used, the characteristics being changed by switching. The purpose of the recording

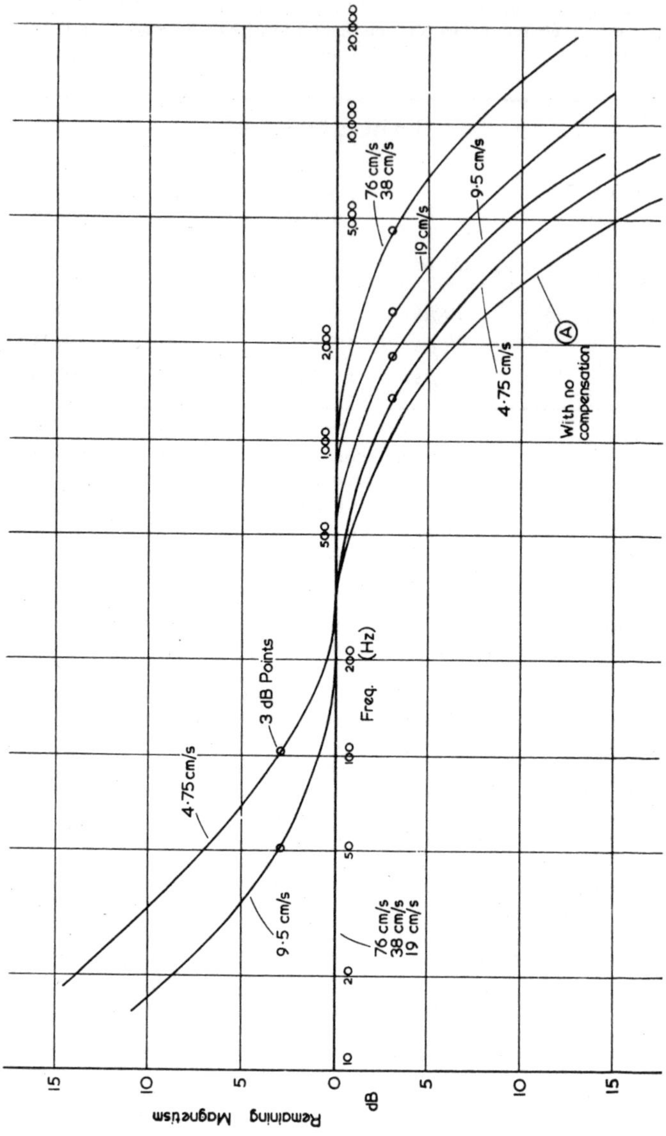

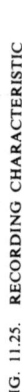

FIG. 11.25. RECORDING CHARACTERISTIC

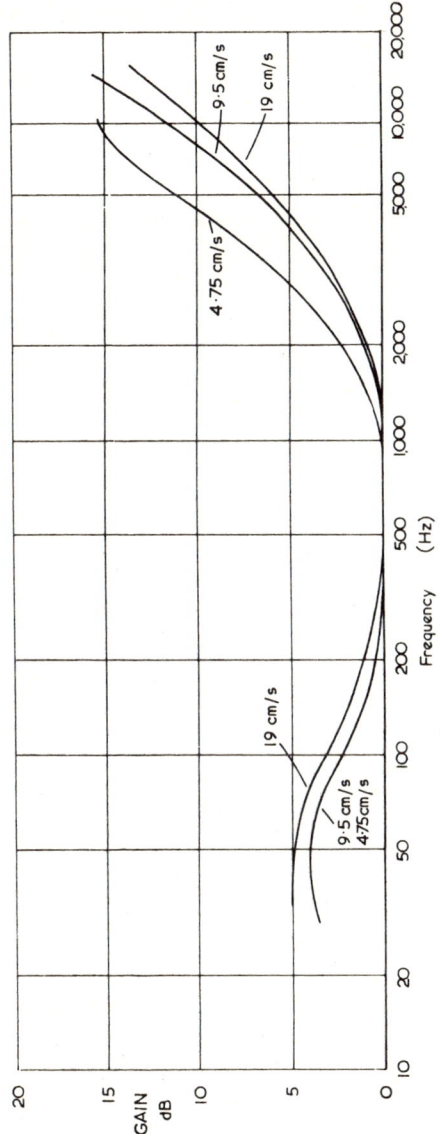

FIG. 11.26. TYPICAL RECORDING AMPLIFIER CHARACTERISTIC

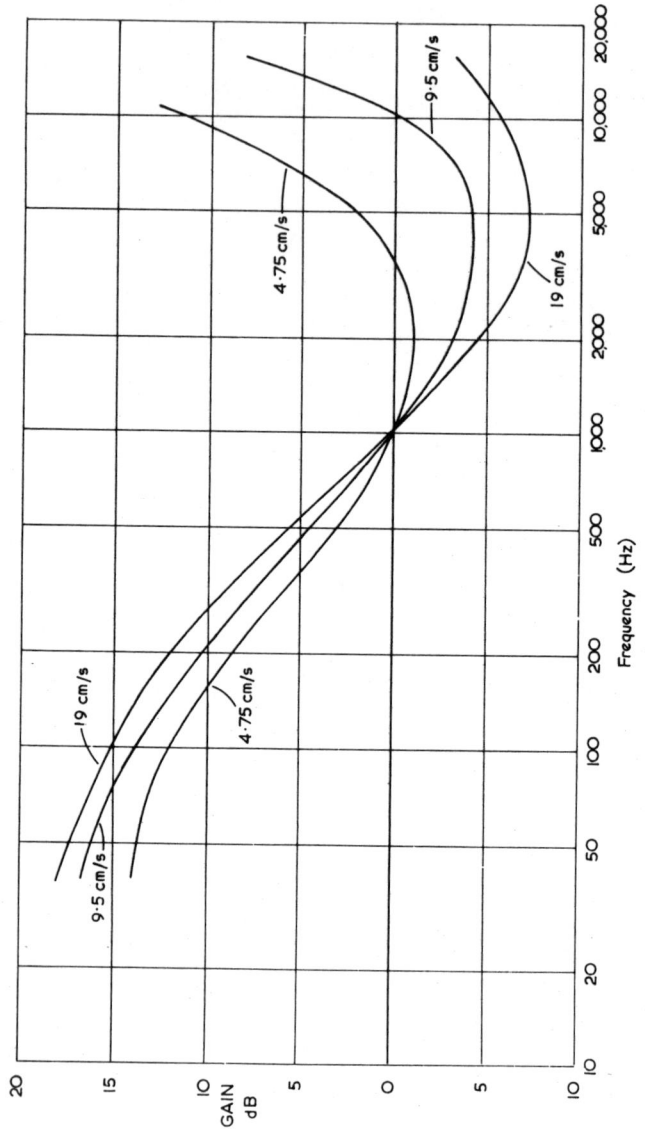

FIG. 11.27. TYPICAL REPLAY AMPLIFIER CHARACTERISTIC

amplifier is to increase the signal input to a sufficiently high value to send the required current through the head, and also provide the required equalization. A block diagram of a common arrangement is given in figure 11.28. The sensitivity of the preamplifier

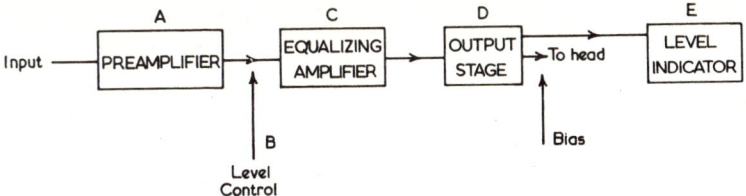

FIG. 11.28. BLOCK DIAGRAM OF RECORDING SECTION OF TAPE RECORDER

is commonly about 0·2 mV, so a microphone can be used; the voltage required to feed the head will be, say, 2 volts. Thus the overall gain required is

$$\frac{2}{0\cdot0002} = 10{,}000 \text{ times}$$

or 80 dB. The input is fed to the preamplifier A, where the voltage is increased to a reasonable value before application to the level control, the position of which is a compromise. If it is placed across the input it will decrease the signal level (except in the maximum position) and hence will reduce the signal-to-noise ratio. If it is placed late in the amplifier then the preamplifier may be too easily overloaded. Most tape recorders are intended to be used with low impedance (600 Ω) microphones and the input impedance is 1 to 2 kΩ, which fits well with these microphones. This input is not likely to be overloaded by a microphone, but one often wants to feed the tape recorder from a radio or another tape recorder. The microphone input is likely to be overloaded with an input greater than, say, 20 mV. In order that larger signals may be applied a potential divider circuit as in figure 11.29 is commonly used. The low-level input connection (sometimes called radio input) will have a sensitivity of 1 to 5 mV with an

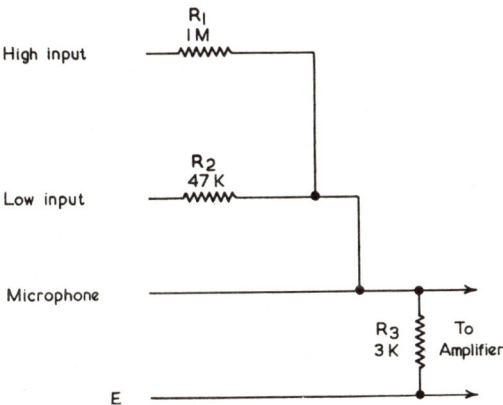

FIG. 11.29. USE OF POTENTIAL DIVIDER ON INPUT

input impedance of 50 kΩ, but will overload with inputs exceeding 100 to 500 mV. Accordingly a high-level input socket is provided (sometimes called the line input), which has a sensitivity of 50 to 100 mV and an input impedance of 100 kΩ to 1 MΩ. The maximum input that can be applied without overloading is 4 to 20 volts. Unfortunately, these input sensitivities are not the same for all tape recorders. The use of a potential divider in this way is not ideal because a large signal is reduced to a small one, which tends to reduce the signal-to-noise ratio. If an input level is being used which requires the level control to be almost fully advanced, then the noise will be high; a more sensitive input socket should be used. It is better if the preamplifier is cut out of circuit when a large signal is available, which is done on some tape recorders.

After passing through the level control the signal goes through the equalizing amplifier C, which has the required frequency response. This now feeds the output stage D, which drives the recording head, whicn also has a suitable bias voltage applied to it. So that the level of the recording signal can be monitored a level indicator E is usually connected to the output stage.

Turning now to more details of the blocks, it is obvious that very many circuits are possible and those shown can only be examples. Figure 11.30 shows the preamplifier of a small battery operated tape recorder. This is a straightforward amplifier with R_2

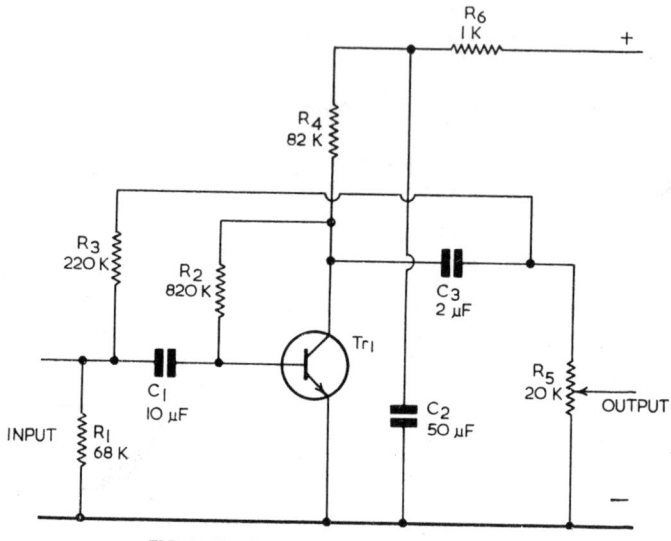

FIG. 11.30. TYPICAL PREAMPLIFIER CIRCUIT

providing base bias, but, since it is connected to the collector, there will be both d.c. and a.c. negative feedback. C_1 and C_3 are d.c. blocking capacitors and the collector load is R_4, R_6 and C_2 being decoupling. R_3 provides negative feedback from the output to the input to improve the linearity. R_5 is the level control. Much more complex circuits may be used depending on whether the tape recorder is mains or battery operated and on its cost.

The circuit of an equalizing amplifier is given in figure 11.31, this being taken from a mains-operated recorder. This also incorporates the ouput stage. The three transistors are directly coupled, Tr_3 acting as an emitter-follower. D.C. and a.c. feedback between Tr_1 and Tr_2 is by resistor R_4, this reducing the gain to about 100. This type of direct-coupled amplifier is commonly used and may be used for the preamplifier. R_1 is the load for Tr_1 and R_5 the load for Tr_2. Turning now to the selective feedback network, at middle and low frequencies the reactance of C_2 will be high (8 kΩ at 1000 Hz) and therefore feedback is through R_{10} and C_1, C_1 having a low reactance at middle and high frequencies (800 Ω at 1000 Hz). At high frequencies the reactance of C_2 will decrease (800 Ω at 10 kHz) and so reduce the feedback and increase the gain as required. The reactances of C_3 and C_4 will also decrease and reduce the feedback (1600 Ω at 10 kHz). The reactance of C_1 will increase at low frequencies (8 kΩ at 100 Hz) and reduce the feedback, so giving the increased gain required at low frequencies. When the recorder is designed for a number of tape speeds then some of these compen-

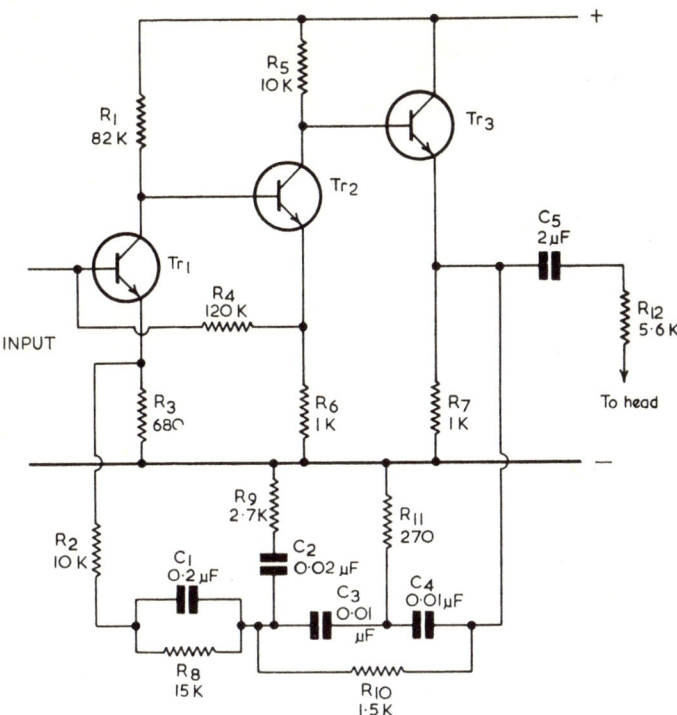

FIG. 11.31. TYPICAL RECORDING EQUALIZING AMPLIFIER

sating components must be changed for different speeds. C_5 acts as a d.c. blocking capacitor.

It is important to note that at all times reference has been made to the recording CURRENT not voltage. It is, of course, the current that matters, since it produces the magnetic field and magnetises the tape. Since the record head is inductive, for a constant current, the voltage across it will rise with frequency. To prevent the current decreasing with frequency the head should be fed from a constant current source, and in this case this is done by the inclusion of R_{12}, which should be high compared with the reactance of the head at high frequencies. In some cases a transistor amplifier is used which has a high output impedence, in which case R_{12} is not required. In some recorders an ouput stage similar to that used for a.f. output amplifiers is used; this amplifier may be used as the output stage on replay.

The recording head must, of course, also be fed with a bias voltage from the bias oscillator, and one usual arrangement is shown in figure 11.32. Since the two voltage sources are in parallel one wants to prevent the oscillator feeding into the amplifier and also prevent the signal being fed into the oscillator. The head is fed through R_{12} (the same R_{12} as in figure 11.31) and a tuned circuit $C_1 L_1$. This is a rejector circuit tuned to the bias frequency and therefore prevents the bias voltage being fed into the amplifier. The head is fed from the oscillator through C_2, which is made fairly small so that little a.f. voltage is fed into the oscillator. The magnitude of the bias is adjusted by R_1. Instead of R_1, preset capacitors may be used. The ratio of bias voltage to signal voltage will be higher than the ratio of bias current to signal current because of the inductance of the head. In practice it is quite difficult to see the signal voltage across the head owing to the large bias voltage.

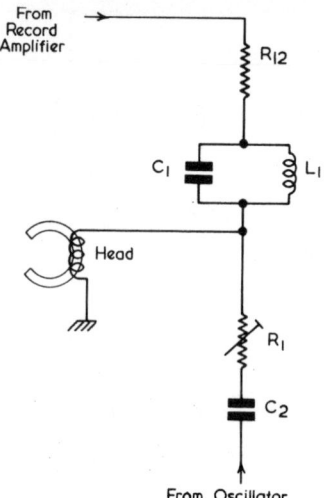

FIG. 11.32. METHOD OF FEEDING BIAS TO RECORDING HEAD

LEVEL INDICATOR

It is necessary to be able to monitor the level of the recording because if it is too large, distortion will occur due to saturation of the tape; if it is too small the signal-to-noise ratio will be poor, because the noise remains approximately constant independent of recording amplitude. In most recorders a small meter is used for this purpose, and in cheap recorders this may be just marked in red and green sections. If the needle goes into the red section then overloading and distortion will result. In battery recorders the same meter is often used to check the battery voltage.

In better domestic recorders what is called a VU (volume units meter) is fitted, calibrated in dB and having a scale as in figure 11.33. The normal maximum reading is 0 dB, and if the needle goes beyond this (often marked in red) then overloading will

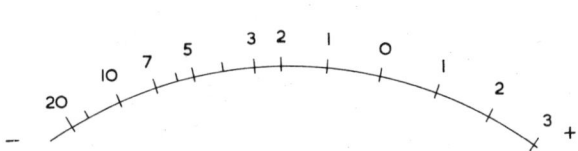

FIG. 11.33. VU METER SCALE

occur. When recording a normal signal (speech or music) the meter reading will be continually changing, the meter moving up and down the scale according to the average volume. For most of the time it will be much less than 0 dB. Two circuits for a VU meter are shown in figure 11.34. At (a) the input voltage is rectified by the diode which charges up the capacitor C. The sensitivity can be varied by R_1, which will also determine the rate of rise of the meter. The rate of fall will depend on the value of C and the meter resistance (and the properties of the meter itself). At (b) R_1 and R_3 supply a small bias but not sufficient for any appreciable emitter current to flow. The signal will cause Tr_1 to conduct on positive half-cycles and so give a meter reading. R will control the sensitivity. C smoothes the half-cycle across the meter and also influences the rate of fall of the meter reading.

A VU meter responds to more or less the average signal, and if a short duration high value signal occurs then the meter will not indicate this correctly and distortion

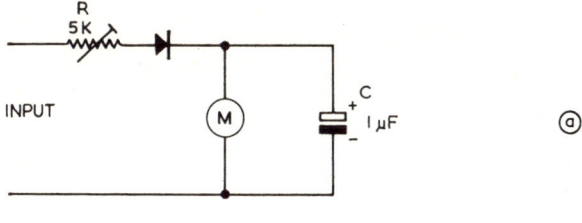

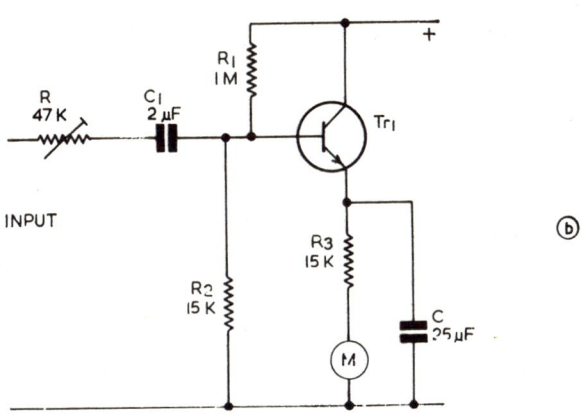

FIG. 11.34. VU METER CIRCUITS

may occur. Professional recorders therefore use a peak programme meter (PPM) and, as its name implies, this is made to read the peak value of the signal. It is made so that the reading rises very quickly but, by using a large capacitor, the reading falls slowly. Some prefer the VU meter and others the PPM.

In figure 11.28 the level indicator was shown connected across the output of the output stage (*i.e.* across the head and series resistor) and this is the common arrangement. It therefore gives a reading proportional to the recording current. Thus, if a constant input voltage is fed to the recorder the reading of the level indicator will rise with frequency due to the equalization given in the equalizing stage. Placing the meter in this position prevents overloading of the output stage and is generally satisfactory. In some domestic and in professional recorders it is placed before the equalizing amplifier, but in this case the output stage must be designed so that overloading does not occur at high frequencies. For this reason the frequency response of a tape recorder MUST be measured at a recording level not greater than −20 dB or most misleading results will be obtained.

REPLAY AMPLIFIER

A block diagram of the arrangements for replay are given in figure 11.35. This consists of an equalizing amplifier followed by an output stage, which is often an emitter-follower designed to feed the preamplifier of an external main amplifier if a tape deck, or the internal amplifier when fitted. The latter will not be considered as this is a normal amplifier already described in previous chapters. An equalizing amplifier circuit is given in figure 11.36. This is a direct-coupled amplifier with Tr_3 as an emitter-follower. D.C. feedback and bias for Tr_1 is from the emitter circuit of Tr_2. The selective negative feedback circuit is R_7, R_8 and C_3. At low frequencies the reactances of C_3 will be high

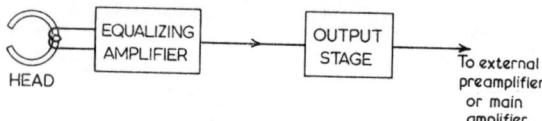

FIG. 11.35. BLOCK DIAGRAM OF REPLAY SECTION OF TAPE RECORDER

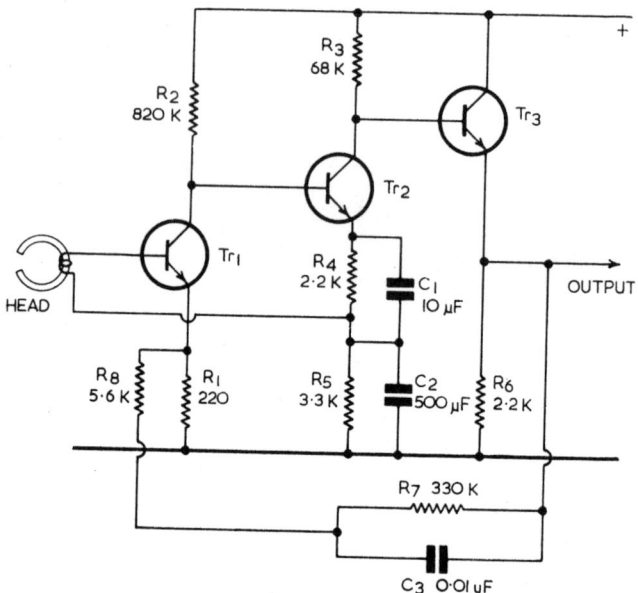

FIG. 11.36. A REPLAY EQUALIZING AMPLIFIER

(330 kΩ at 50 Hz) and hence the amount of feedback is mainly settled by R_7 and R_8 and will be small, hence the gain is large as is required. As the frequency increases, the reactance of C_3 will decrease (1·6 kΩ at 10 kHz), but when it becomes comparable with the value of R_8 it will not make much difference to the feedback. Thus the feedback increases as the frequency increases, but reaches an approximately constant value at, say, 2 to 3 kHz. Therefore, the gain is reduced at first but then becomes constant. There is no arrangement in this circuit to give the high frequency boost required, but this can be obtained in the head design, and particularly by using a capacitor across the head. This capacitor causes resonance to occur with the reactance of the head at high frequencies. The gain at mid frequencies is about 200.

A possible output stage is given in figure 11.37, having a gain of about 10 times, being controlled largely by the ratio of R_3 to R_4. A preset volume control may be placed between the equalizing amplifier and output stage, so that the output level can be set to the correct value. There does not seem to be a standard level but it is commonly about 1 volt with an output impedance of 10–15 kΩ or less. When a three-head machine is used a higher inductance head can be used, which will produce a higher voltage and hence less gain is required.

AUTOMATIC LEVEL CONTROL

Up to now it has been assumed that the control of the record level is manual. However, many recorders are now fitted with automatic level control systems. The idea is that the louder the signal the less the gain, so that the recording level is approximately independent of the signal input level. The arrangement is convenient but is a compro-

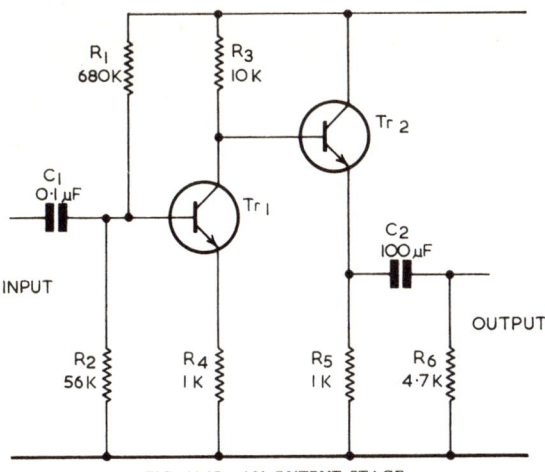

FIG. 11.37. AN OUTPUT STAGE

mise, as obviously all parts of a recording do not want to be at the same level. The speed of response of the circuit is important. It should act as rapidly as possible on increasing signals, so that overloading does not occur on the sudden arrival of a loud passage. the speed at which the gain increases should be slow, say 20–60 seconds. If this is too short then the gain increases between words on speech and can badly distort music. For example, with a piano note should die away, but if the gain increases too rapidly this dying away is prevented. Sometimes a shorter time is used on speech than on music. Automatic level control or automatic gain control certainly prevents over-loading and avoids the need to watch the level indicator. However, it can never be perfect and it is not suitable for all occasions. Some recorders do not provide a level indicator, but it is much better to have one, even with automatic level control, other-wise there is no method of checking that recording is actually taking place or whether the automatic level control circuit is working. In some recorders manual or auto level control can be selected by a switch.

As in any automatic control system we require a control signal, which in this case is obtained from the output stage, and a device which will control the gain. The common device used is a diode or transistor, which forms a potential divider circuit early in the recording amplifier. A typical arrangement is shown in figure 11.38, R_3 and D_3 forming this potential divider. The output from the recording amplifier is fed through a filter $R_1 C_1$ to remove any bias voltage. The signal is then rectified by D_1 to produce a voltage across $R_2 C_2$ proportional to the signal amplitude. The voltage across C_2 is fed to D_3, through diode D_2. Thus, the greater the signal level, the greater the voltage applied to D_3. If the I-V characteristics of a diode are examined [as at (b)] it is seen that the greater the voltage the less the slope or dynamic resistance of the diode. Thus, when there is a large signal the voltage across the diode may correspond to C where the slope resistance is low. R_3 and the diode form a potential divider so that the signal to the equalizing amplifier is reduced, so reducing the record level until equilibrium is reached. The rate of rise of voltage across C_2 is settled by the resistance of D_1, together with R_1 and the output impedance of the recording amplifier. These resistances are small, so the rise time is short. The rate of decrease of voltage across C_2 is settled by R_2 and the current in D_2 and D_3, and these are of such values that the rate of fall is slow. D_2 is necessary to prevent shorting the signal to earthy line by C_2. Since the diode characteristic is curved, this circuit must be used where the signal level is low, say a few millivolts, or distortion will occur.

Circuits using a transistor in place of a diode are given in figure 11.39. At (a) the transistor (which has no d.c. applied to its collector) is connected through the d.c.

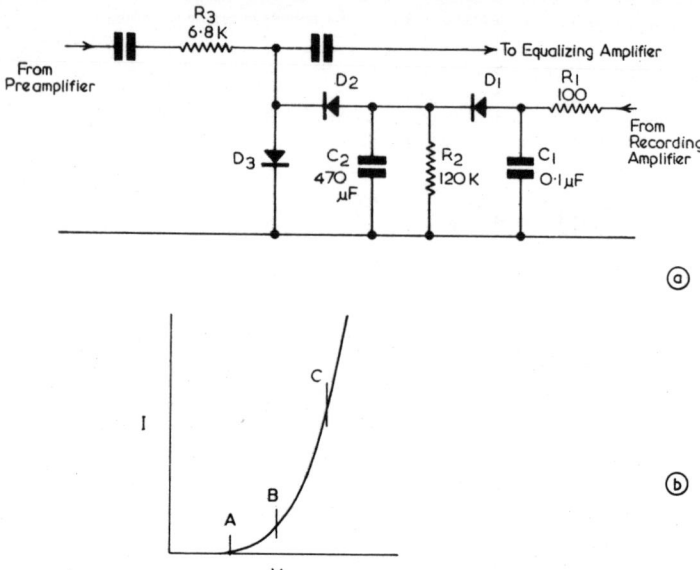

FIG. 11.38. (a) AUTOMATIC LEVEL CONTROL CIRCUIT. (b) DIODE CHARACTERISTIC

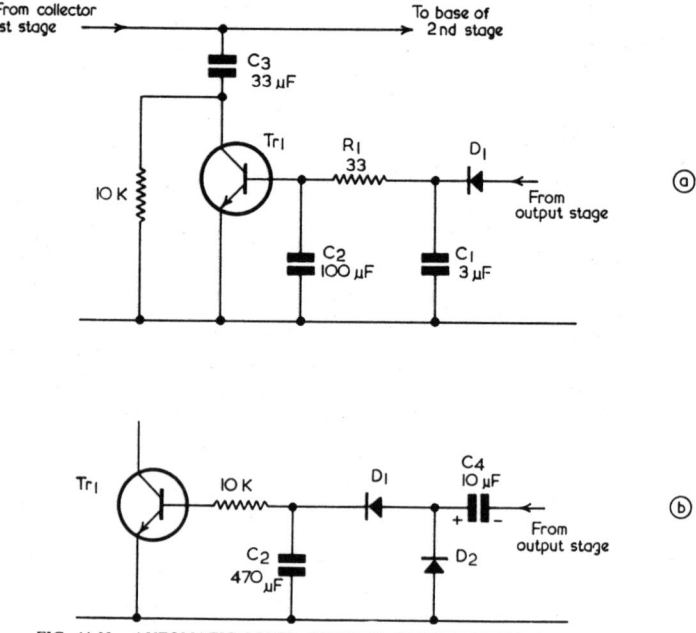

FIG. 11.39.   AUTOMATIC LEVEL CONTROL CIRCUIT USING TRANSISTOR

blocking capacitor, so as to form a potential divider with the output impedance of the first stage. The effective resistance of the transistor is decreased by applying a voltage to its base. The voltage from the output stage is rectified by D_1 and charges up C_1 and C_2, so controlling the transistor. R_1 controls the rate of rise of voltage on C_2 and the rate of fall is now settled by the base current in Tr_1. Figure 11.39(b) is a similar circuit except that a voltage doubler rectifier circuit is used.

A field effect transistor (FET) may be used in place of the diode or bipolar transistor, and a circuit is given in figure 11.40. The source connection of the FET is given a positive voltage from the zener diode D_1, which is fed through R_1 from the supply.

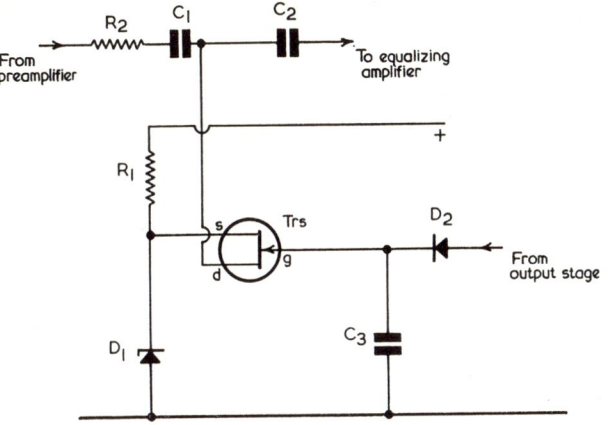

FIG. 11.40. AUTOMATIC LEVEL CONTROL CIRCUIT USING F.E.T.

With no voltage across C_3, the FET acts as almost an open circuit and there is no reduction in signal value between the amplifiers. The voltage from the output stage is rectified by D_2, C_2 becomes charged so making the gate positive with respect to the earthy line. This reduces the effective resistance of the FET. R_2 and the FET form a potential divider, so reducing the voltage fed to the equalizing amplifier. An FET is less likely to introduce distortion than a diode or bipolar transistor.

LIMITER CIRCUITS

The characteristic of an automatic level control circuit is given in figure 11.41. As the input level increases the gain is reduced. A disadvantage is that there is no manual control over the recording. Difficulties do arise if good recordings are required, one of which is that if the signal level is very small the recording amplifier sets itself to maximum gain and therefore the noise level becomes high. As soon as a signal of reasonable value occurs the gain is reduced to normal, but this high noise level at the start may not be acceptable. Even a short click will cause the gain to drop.

It is better to use what is called a "limiter". This rather implies a circuit which is

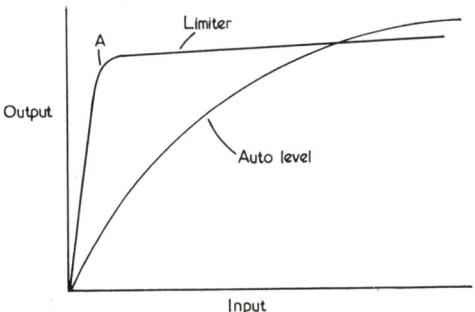

FIG. 11.41. CHARACTERISTICS OF LEVEL CONTROL CIRCUIT AND LIMITER CIRCUIT

going to clip or limit the peaks of the signal and so cause distortion, but this is not
the case. In practice this system produces very little distortion and even then only
when overloading would normally take place. A typical characteristic is shown in
figure 11.41. Up to point A, which corresponds to 0 dB recording level, the limiter
has no effect and the recording level can be set to any required value manually. If the
signal increases beyond point A the limiter comes into operation and prevents
appreciable increase in recording level, as shown. For best operation the recording
level should be set manually (with the limiter switched off) until the needle just reaches
the 0 dB level on loud passages and then the limiter switched on. If any unsuspected
louder passages now occur the limiter will take care of them. The same basic circuits
as for automatic level control can be used, it only being necessary to prevent the
circuit operating for recording levels below a certain value. For example, in the
circuit of figure 11.40, if the voltage of D_1 is increased it will prevent the circuit
operating below a certain value.

OSCILLATORS

The oscillator should produce as good a waveform as possible and should be
free of second harmonic distortion. Considerable power, 1 to 3 watts, is required.
A typical circuit is given in figure 11.42, where the erase head is made the tuned circuit
by adding C_1 across it. Coupling from the collector circuit to the base circuit is by the

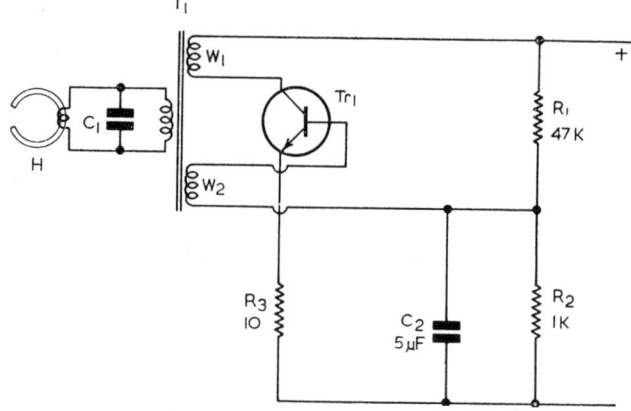

FIG. 11.42. AN OSCILLATOR CIRCUIT

windings W_1 and W_2 on the transformer. To make certain that oscillation will start
a small forward bias is provided by $R_1 R_2$, and R_3 limits the flow of emitter current so
reducing distortion Another circuit is given in figure 11.43, this being a Colpitt's
oscillator circuit. The erase head H and C_1 to C_4 form the tuned circuit. R_1 provides
some forward bias to start oscillation. L_1 offers a high impedance in the emitter
circuit and allows an a.c. voltage on the emitter, as is essential in this circuit since the
collector is at positive line potential. This type of circuit is commonly used on small
domestic tape recorders, the oscillator also, of course, supplying the bias voltage to
the record head.

On some tape recorders the a.f. output stage is used as the oscillator on record, it
not being required as an output stage on record. A circuit is shown in figure 11.44.
Tr_1 and Tr_2 form the normal complementary transistor output stage and they feed
the erase head through C_5 (d.c. blocking capacitor) and transformer T_1. This transformer is tuned by C_3 and the erase head. The bases are fed through C_1 and C_2 from
transformer T_1. Since Tr_1 and Tr_2 are connected as emitter-followers they have no
voltage gain, but the circuit can be made to oscillate by using the step-up transformer

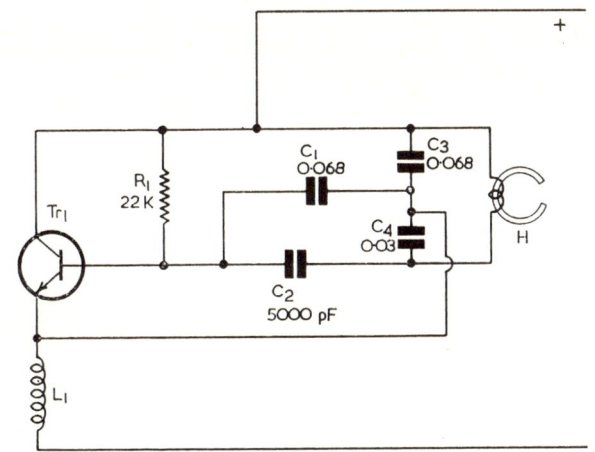

FIG. 11.43. COLPITT'S OSCILLATOR CIRCUIT

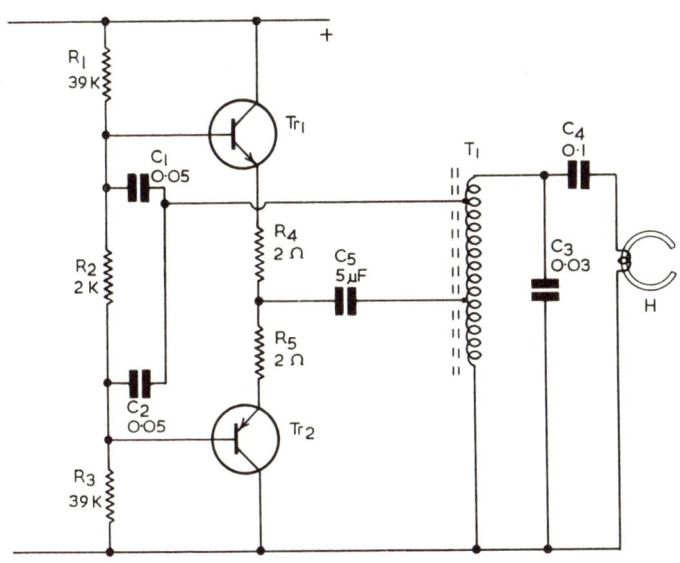

FIG. 11.44. OUTPUT STAGE AS OSCILLATOR

T_1. R_2 provides bias to Tr_1 and Tr_2 and since this is a push-pull circuit any distortion should be small.

In the more expensive recorders a more complex arrangement may be used, and one is shown in figure 11.45. The oscillator is now followed by a power amplifier, so that the heads do not influence the oscillator. Tr_1 and Tr_2 form the push-pull oscillator with tuned circuit T_1C_3. Feedback is through C_1 and C_2 which with R_1 and R_2, form self-bias circuits. R_1 and R_2 also provide a starting bias. Tr_3, Tr_4 and Tr_5 form a normal complementary output stage, Tr_3 being the driver stage which is fed from the secondary winding on T_1. R_5 and R_6 provide bias for Tr_3, and D_1 bias for Tr_4 and Tr_5. R_7 is the load resistor on Tr_3. Many other oscillator circuits are possible.

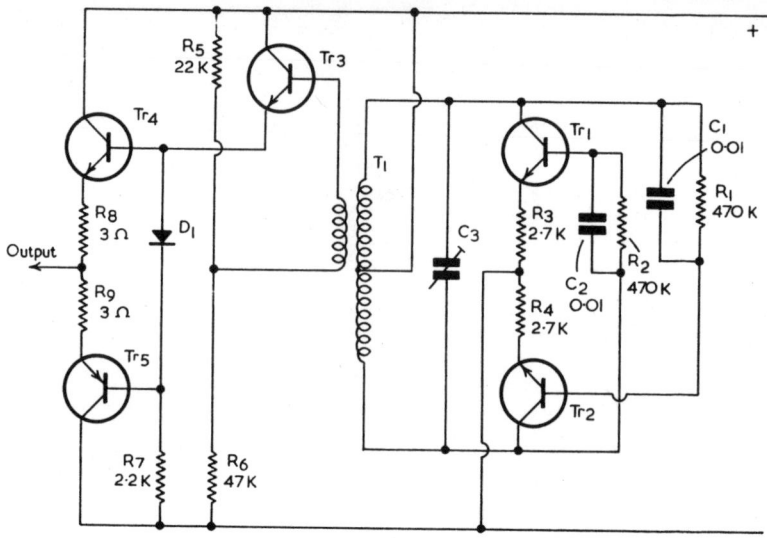

FIG. 11.45. PUSH-PULL OSCILLATOR AND AMPLIFIER

MECHANICAL DETAILS

Only the general principles will be considered, as the detailed mechanical arrangements are numerous and varied. Perhaps the most important part of the tape recorder
is the capstan and the pressure roller which keeps the tape in contact with it, as these
two components determine the speed of the tape. Theoretically the tape should be
driven only by the capstan, and this will occur if the pressure roller is made of a width
less than the tape. However, this requires a large pressure between capstan and tape
to prevent the tape slipping. This can be overcome by making the pressure roller
wider than the tape so that the roller makes contact with the capstan on each side of
the tape. The capstan now drives the pressure roller and the roller drives the tape. Thus
the tape is partly driven by the capstan and partly by the pressure roller. To get good
results the pressure roller must be very uniform and of suitable material. If wow and
flutter are to be avoided the capstan must be made very accurately. It must be per-
fectly round and have no eccentricity, and of course must run in good bearings. The
tape is pulled over the heads by the capstan, and the tape must be maintained in
contact with the heads. In some cases the tape is pressed against the heads by felt pads.
Greater pressure will ensure better contact, but more wear on the heads. In other cases
the pressure is maintained by a suitable back pressure from the supply reel.

When the tape has passed the capstan it must be wound on the take-up spool.
The speed of the spool will vary, being much slower when it is full. For a given shaft
torque the pull on the tape will also vary, being a minimum when the spool is full.
The minimum pull must be sufficient to wind the tape reasonably tightly on the spool
and the maximum pull must not be such as to upset the speed of the capstan. On
expensive recorders some type of tension control may be used, but this is not common.
The take-up spool may be driven from the same motor as drives the capstan, usually
through belts or rollers and a slipping clutch. In other cases a separate motor may be
used, run at reduced voltage for spooling on record or replay.

Similarly, the supply spool must have some back tension to maintain the tape
taut in order to maintain the tape in contact with the heads. This may be done by a
slipping clutch or a separate motor run at reduced voltage. Tape recorders are also
provided with fast forward and reverse winding facilities. Under these conditions

the pressure roller is not engaged with the capstan, and the tape is lifted off the heads to reduce wear. The tape is now driven by one of the spools, a suitable reverse tension being applied to the other. They may be driven from the capstan motor or by separate motors. When separate motors are used one is supplied with full voltage and the other with reduced voltage (in reverse), so that a suitable tension is maintained on the tape. A good braking system is essential to prevent spillage of tape on stopping, it being essential to brake both supply and take-up spools—particularly important on fast wind or re-wind. The brakes are commonly friction brakes using felt pads; various arrangements are used.

The mechanical arrangements to perform these functions are complex, particularly with a single motor. The controls may be of the "joy stick" type or press buttons. Ganged to these controls are the necessary switches to change for record mode to replay mode. In some recorders the operations are performed by solenoids, energized by press buttons or switches. This tends to simplify the mechanical design and allows for remote control operation.

Some recorders are fitted with a pause button. This disengages the pressure roller or idler wheel from the capstan and either removes the drive from the take-up spool or clamps the tape. This enables the tape to be stopped and started almost instantly, and is very useful for editing. When the normal start control is used the tape takes a short time to reach its correct speed.

DRIVE MOTORS

We will now consider the motors used to drive the tape recorder. In a mains recorder it is normally an induction motor or hysteresis motor. The induction and hysteresis motor depends on the production of a rotating magnetic field. In large induction motors this is obtained by three windings connected to a three-phase supply, and in this case a uniform rotating field is easily obtained. With a single-phase supply the ideal way of producing a rotating field is to use two sets of coils at right angles to each other and feed the two coils with currents having a phase difference of 90°. Under these conditions it can be shown that a rotating field is produced, this rotating at synchronous speed, *i.e.* 50 revolutions per second, or 3000 r.p.m. for a 50 Hz supply. There are two ways of producing a rotating field in small motors fitted to tape recorders.

(a) Use of capacitor

This is shown diagramatically in figure 11.46 where two sets of coils C_1 and C_2 are used, coils C_1 being fed directly while C_2 are fed through a capacitor C. Due to the

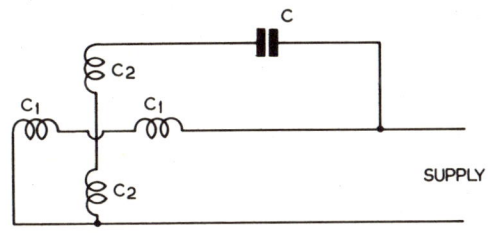

FIG. 11.46. CAPACITOR TYPE MOTOR

capacitor the current in C_2 windings will lead the current in the C_1 windings, the phase angle being approximately 90°. Thus a rotating field is produced as required. In order to keep the reluctance of the magnetic circuit to a minimum the coils are wound in slots in a laminated core or stator S as in figure 11.47. The rotor R is iron, so that the magnetic circuit is all iron, apart from the small air-gap. This type of motor is more expensive than the shaded pole motor to be described; but it is a more ideal arrangement, producing a more uniform torque and less external field.

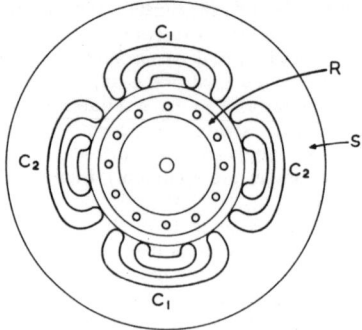

FIG. 11.47. CONSTRUCTION OF INDUCTION OR HYSTERESIS MOTOR

(b) Shaded pole motor

One method commonly used in constructing this type of motor is shown in figure 11.48, consisting of a laminated stator S. On the stator is the coil C, which produces the magnetic field. R is the rotor. The pole-pieces P are now divided into

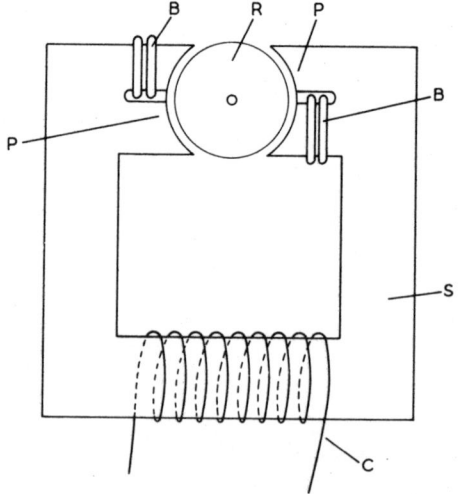

FIG. 11.48. SHADED POLE INDUCTION MOTOR

two parts, one part having copper shading rings B. These are closed rings and the flux passing through them causes currents to be induced in them. It can be shown that as a result of these currents the flux passing through the rings lags in phase with the flux in the rest of the pole. The angle is not 90° but, say, 30° to 40°, which is sufficient to cause a crude rotating field. This method of construction is cheap and very often used.

We must now consider the rotor, first that used on an induction motor. A typical, so-called cage rotor is shown in figure 11.49 consists of a laminated cylinder C with holes or slots in which are placed copper bars B. The bars are all joined together at the ends by the end rings R. The rotating field will cause an e.m.f. to be induced in the bars, and hence a current flows. This current causes a force to be produced in such a direction as to reduce the e.m.f. (Lenz's law), which means that the force is in the

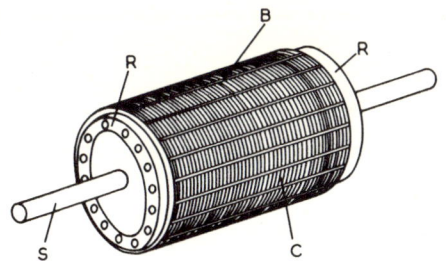

FIG. 11.49. ROTOR OF CAGE TYPE INDUCTION MOTOR

direction of rotation of the field. Thus the rotor will rotate, so that the rate at which the field cuts the bars is reduced and hence the current is reduced. The rotor will run up to almost the speed of the rotating field, but must always be slightly less, otherwise there would be no e.m.f. induced in the bars and neither current nor force would be produced. The difference between that of the field (synchronous speed) and the actual speed is known as the "slip" and increases as the load is increased. The amount is only a small percentage. Thus the induction motors run at an almost constant speed provided the frequency is constant (which determines the synchronous speed). This type of rotor is used in both types of stator (figures 11.47 and 11.48). The induction motor is the one most commonly used.

The hysteresis motor consists of a rotor of hard steel or a magnetic material such as Alnico. The rotating field now produces a torque on the rotor, due to the large hysteresis in the rotor, and the speed now increases until the rotor runs at the SAME speed as the field, i.e. it runs synchronously. Provided the frequency is constant this type of motor runs at an exactly constant speed, and is more expensive than the induction motor. It is normally used only with the construction of figure 11.47.

If the machine has only two poles, as described, then the synchronous speed is 3000 r.p.m. (for a 50 Hz supply), but if four poles are used the synchronous speed becomes 1500 r.p.m.

Battery-operated machines must operate from the low voltage d.c. supply. A normal d.c. motor is commonly used with a permanent magnetic field. In some recorders no speed control is used, but it is common practice to control the speed electronically. One common circuit is shown in figure 11.50, the component values being only approximate varying with different machines. The motor is fed through the main control transistor Tr_1, and resistor R_5 of low value. The actual voltage on the motor may be only 4 volts, the remainder being dropped across R_5 and Tr_1 (most of it across Tr_1). The transistor Tr_1 is controlled by Tr_2 and the greater the collector current of Tr_2, and hence base current of Tr_1, the less the voltage drop across Tr_1 and the greater the motor voltage. The emitter voltage of Tr_2 is the motor voltage, less the drop across D_1 and D_2; since these are forward biased the drop across them is approximately constant. Neglecting the drop across R_5, the base voltage of Tr_2 is a fraction of the motor voltage. Suppose that the motor voltage rises by, say, 1 volt. Since the voltage across D_1 and D_2 is constant, the emitter voltage will also rise by 1 volt. The base voltage will rise by a smaller amount, because of the potential divider R_1, R_2 and R_3. Thus the emitter goes positive with respect to the base and the current of Tr_2 is reduced. This reduces the base current of Tr_1, so causing the voltage across it to rise and reduce the motor voltage until equilibrium is reached.

Consider now the voltage across R_5 which is applied between base and emitter of Tr_2. If the current of the motor increases then the drop across R_5 increases, so making the base of Tr_2 more positive with respect to the emitter. This increases its collector current, and hence the base current of Tr_1. Also, it reduces the drop across Tr_1 and, if the circuit is designed correctly, this increased voltage applied to the motor will make up for the resistance drop in it due to the increased current, thereby maintaining a

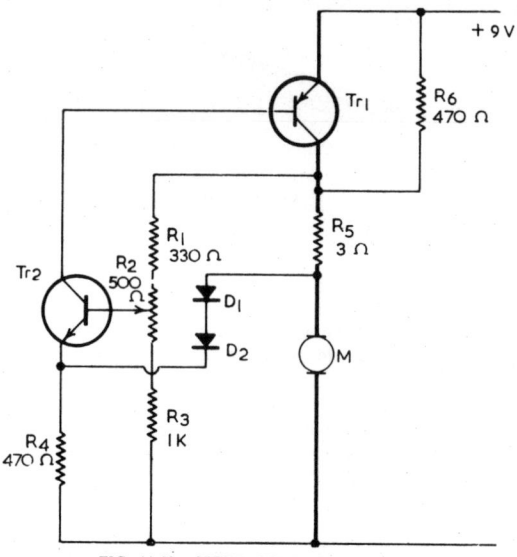

FIG. 11.50. SPEED CONTROL CIRCUIT

constant speed.

In some expensive tape recorders more elaborate arrangements are used, the commutator of the motor being replaced by transistors. These are too involved to be described in this book.

SPECIAL FACILITIES

Sound-on-sound

This is a facility provided on some recorders having three heads. The idea is to be able to mix two programmes, one being previously recorded. The first programme is recorded on one track, say the L track. The recorder is now set so as to play back from the L track and this signal is fed to the R track, together with another signal from, say, a microphone, so that the combined signal is recorded on the R track.

Echo

Owing to the delay between record and playback heads, due to the spacing of the record and replay heads, it is possible to get an echo effect by feeding back some signal from the playback head and mixing it with the material being recorded.

Simul-sync recording

A four CHANNEL recorder is made by TEAC in which each of the four channels can be recorded or played back independently. This may be used for quadraphonic recording (see next chapter). It can also be used to make multitrack recordings, e.g. a recording of a number of instruments played by the same person or a duet by one person. A base recording is made on track 1, then played back and monitored while the next recording (say another instrument) is put on track 2. If the normal playback head is used for track 1 there will be a time displacement between the two recordings because track 1 is being monitored at the position of the playback head, while track 2 is being recorded at the position of the record head. To overcome the time delay problem the recorder is made so that the Track 1 (and any other track) can be monitored off the record head. The quality of playback from the record head is not as good, but this is immaterial as it is only used for monitoring purposes.

CASSETTE RECORDERS

A disadvantage of the reel-to-reel machine is that the tape has to be threaded through the recorder and, before a tape can be removed, it must be run through the machine to one end. This disadvantage is overcome in the cassette recorder and cartridge recorder to be described later. The idea of a cassette is to have the spools built into a case, so that no threading is required and the case, complete with tape, can be removed without running the tape to one end. The construction of the compact cassette (a larger one was originally made) is shown in figure 11.51. The case or cassette is rather a complex plastic moulding and all the details are not shown on this drawing.

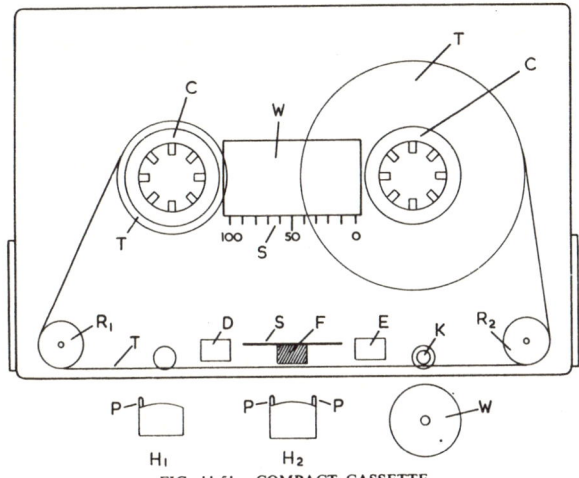

FIG. 11.51. COMPACT CASSETTE

The tape T is wound on the two cores C (spools are not used but only cores, the tape being held in place by the sides of the cassette) and passes over two rollers R_1 and R_2. The core centres have projections so that they can be driven by suitably splined spindles. When the cassette is put into the machine the cores fit over the drive spindles and the cassette is located in position by two pins, which fit into holes D and E. The capstan K of the recorder (about $\frac{1}{16}''$ diameter) fits in a hole in the cassette as shown. When the recorder is operated a pressure roller or idler wheel W presses the tape against the capstan K and the two heads H_1 (erase) and H_2 (record/playback) are moved into place. The tape is pressed against head H_2 by a felt pad F and spring S (part of the cassette). The tension of the tape presses it sufficiently well against the erase head H_1. Pins P on the heads locate the tape axially. It should be noted that the tape on the cassette is wound with the oxide coating outwards, the opposite to that of the reel-to-reel machine.

The cassette can be removed without running the tape to one end. For fast forward or reverse winding the appropriate core is driven and the pressure roller and heads remain in the position shown in figure 11.51. In the centre of the cassette is a window W with a scale marked 0 to 100, so that an approximate indication of the tape position is given. The speed of the tape is 4·75 cm/second ($1\frac{7}{8}''$/second) and because of this slow speed good high frequency response is difficult to obtain. The head H_2 must have an extremely small gap (say 1 micron or 1×10^{-6} metres). Cheaper recorders give a response up to 6000–8000 Hz, but more expensive recorders now give very good results, up to 15 kHz or, with chromium dioxide tape, up to 18 kHz or even 20 kHz. The width of the tape is 3·8 mm (0·15″ or approximately $\frac{5}{32}''$).

The track arrangements for mono recording are given in figure 11.52, where it will be seen that the tape is run in one direction and the BOTTOM track recorded. The cassette is then turned over and the other track recorded. The arrangements for stereo

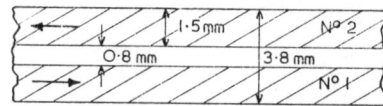

FIG. 11.52. CASSETTE TAPE RECORDING, MONO

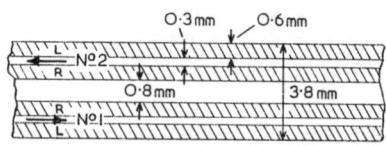

FIG. 11.53. CASSETTE TAPE RECORDING, STEREO

recordings are shown in figure 11.53, where number 1 R and L tracks are first recorded, the tape turned over and number 2 R and L tracks recorded. Since the R and L tracks of the stereo recording are next to each other (not like a four-track reel-to-reel recording) a stereo recording can be played back on a mono machine to produce a mono output. Similarly a mono recording can be played on a stereo machine and will, of course, produce a mono output, *i.e.* the recording is compatible.

The playing time of the cassette depends on how much tape can be wound in the restricted space. Normally three types of cassette are available: C.60, C.90 and C.120. The C.60 plays 30 minutes per side, a total of 60 minutes, and contains about 90 metres (300 feet) of triple play tape 0·018 mm thick. The C.90 plays for a total of 90 minutes and contains 135 metres (450 feet) of quadruple play tape about 0·012 mm thick. The C.120 plays for 120 minutes and contains 170 metres (550 feet) of sextuple play tape 0·009 mm thick. This tape is so thin that it is more likely to cause trouble by jamming in the cassette and cannot be really recommended by the author. There is a shorter tape of 45 metres (150 feet) mainly intended for "letters" and plays for 30 minutes.

It is now fairly common to bring out a recording as a record, as a cassette and often as a cartridge. The normal pre-recorded cassette has, therefore, a playing time the same as an L.P. record, rather less than a C.60. Some pre-recorded cassettes are "doubles" and contain the contents of two L.P. records. Pre-recorded tapes are nearly always stereo recordings.

As well as cassette recorders (which will record and replay tapes) there are cassette players which are ideal for playing pre-recorded tapes. Cassette players are often used in cars and may be combined with a radio. Portable radios are sometimes combined with cassette recorders, so that recordings can be made direct off the radio and, of course, used as a cassette recorder and player.

Most recorders use a single capstan to drive the tape as already explained, but dual capstan recorders are made, which claim to have less wow and flutter. One capstan is run slightly faster than the other so that the tape is kept taut over the heads. One arrangement is shown in figure 11.54, where the two capstans are driven by a single belt. The flywheels are of the same diameter, but due to the stretch of the belt, the take-up flywheel and capstan run slightly faster than the supply one.

When the tape comes to the end it should be stopped. In simple tape recorders it just slips on the capstan, there being a leader of clear plastic which is firmly attached to the core so that it does not become free. The friction of the capstan against the tape causes a squeak, so indicating the end of the tape. In some recorders an automatic stop is provided when the end of the tape is reached. The end is usually detected by the fact that one of the spindles, take-up or supply, has stopped. This may be detected electrically or mechanically. Some cassettes have a strip of foil at the end that is used to detect the end in some recorders.

A useful addition is a digital counter so that a particular portion of the tape can easily be located.

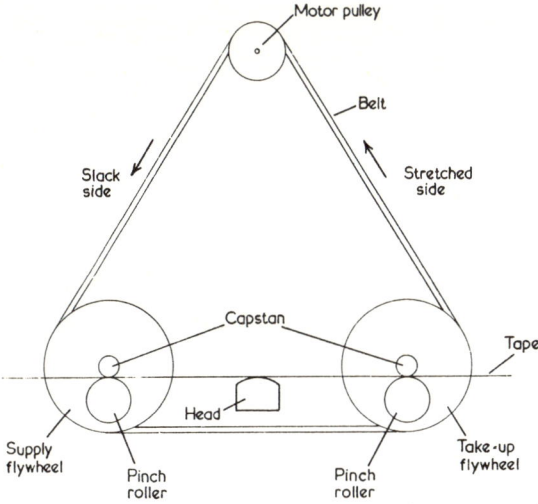

FIG. 11.54. DUAL CAPSTAN ARRANGEMENT

CARTRIDGE RECORDERS

A cartridge differs from a cassette in that a single continuous loop of tape is used on a single spool. The internal arrangements are shown in figure 11.55. The tape used is now $\frac{1}{4}''$ wide and it is run at 9·5 cm/second ($3\frac{3}{4}''$/second). The tape loop T is wound on a spool S (only a one-sided spool) and passes over a guide G after being pulled out of the centre. It then goes over a pressure roller or idler wheel R and to the outside of the spool. Since the diameter at the centre is smaller than the outside, the tape is always kept taut, but it must slide over itself on the spool. It is lubricated so that it

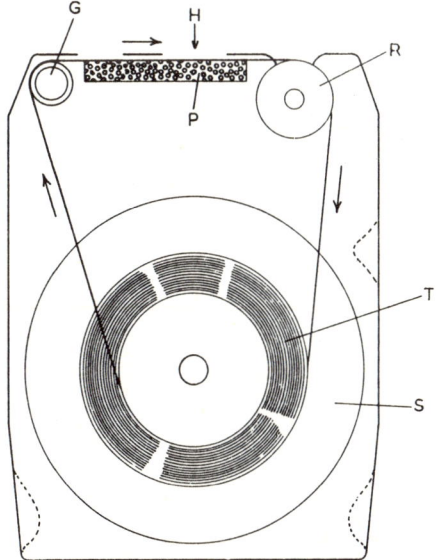

FIG. 11.55. CARTRIDGE

slides freely. The tape is wound oxide out. When in use the cartridge is pushed into the recorder and when in place, the head is at position H and the tape is pressed against the head by the plastic foam P. The roller R also presses the tape against the capstan and the tape is driven by the capstan in the direction shown. Various guides are moulded into the cartridge to keep the tape in the correct position. Due to the use of a single spool in this way the direction of tape traverse CANNOT be reversed, hence there is no fast rewind. The fast forward feature is also often not provided. The tape can ONLY be driven by the capstan, so fast forward drive is obtained by driving the capstan at a higher speed.

The tape uses 8 tracks in stereo, four stereo programmes, and arranged as in figure 11.56. Stereo pairs are now 1 and 5, 2 and 6, etc. and, as far as is known, all recordings are in stereo. In order to change tracks the head is moved physically and a

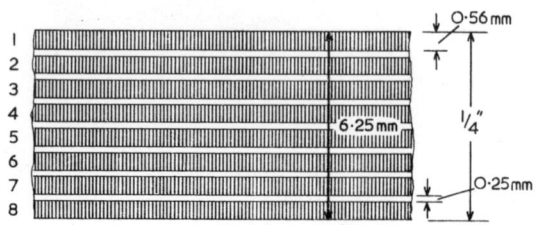

FIG. 11.56. CARTRIDGE RECORDING

control is fitted to switch from track to track. The normal arrangement is a solenoid which ratchets round a cam. Each ratchet position moves the head up one track and, after the fourth, returns it to the bottom track. There is, of course, a joint in the tape which has a conducting foil section. Two contacts by the head detect this foil and may change the player automatically to the next track, so that the whole tape is played and then switched off, or may continue to play the tape continuously until the player is switched off. Alternatively, it may be arranged to stop the player at the end of one track. As in cassettes, the normal pre-recorded cartridge plays the same time as an L.P. record, about 40 minutes. Double ones are available playing two L.P. record programmes. Most cartridge equipments are players only. A few recorders are available but recording is difficult because there is no easy way of knowing the position of the tape, and rewinding is impossible. Cartridge players are most commonly used in cars. Owing to the higher tape speed the high frequency response should be better than that of a cassette, but this does not appear to be the case. Because of the use of lubricant the heads need frequent cleaning. Cartridges are much bigger than cassettes and pre-recorded ones are dearer. Blank cartridges are available in total playing times of 20, 40 and 80 minutes.

NOISE REDUCTION SYSTEMS

One disadvantage of tape recording is the limited signal-to-noise ratio, particularly on cassettes. A number of attempts have been made to increase this and two methods will be described briefly.

(a) Dolby System

This was developed by Dolby Laboratories Inc. and there are two systems: Dolby A and Dolby B. Dolby A is used extensively by recording companies to obtain very high signal-to-noise ratios on reel-to-reel machines. Dolby B is a simplified scheme suitable for domestic equipment, and will be the only one to be described. In some ways it is a kind of pre-emphasis and de-emphasis circuit. It only operates with signals having a frequency above 2 kHz, and below a certain value, say -20 dB. The principle on recording and playback is shown in figure 11.57. The low frequency signal is not altered in any way, but the small amplitude high frequency signal is

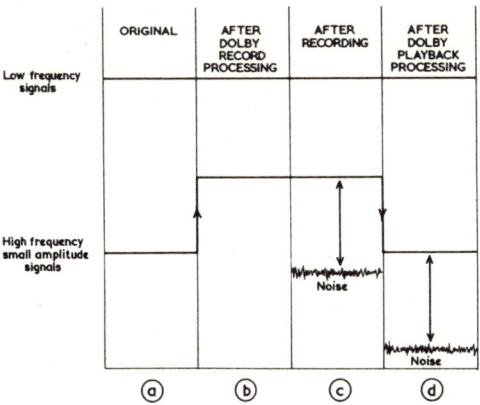

FIG. 11.57. DOLBY RECORDING AND PLAYBACK

increased in value in the Dolby processor, as seen in the figure. After recording there will be some noise present, so many dB below the high frequency signal recorded. After playing back through the processor, the signal level is reduced to the original value and, in so doing, the noise will be reduced by the same amount. If this is compared with non-Dolby recording, as in figure 11.58, it will be seen that the noise is of much higher value in relation to the signal. This is because the recorded noise level is the same during recording, but the signal level is not increased as in the Dolby system.

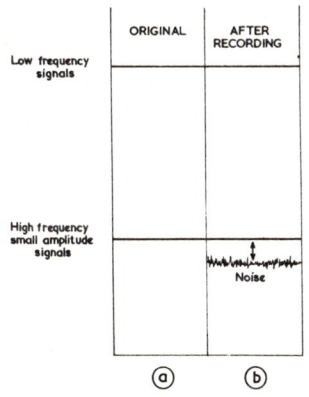

FIG. 11.58. NORMAL RECORDING AND PLAYBACK

The important point about the Dolby system is that it only operates on frequencies above 2 kHz, and even then only on those of small magnitude. Any variable gain circuit tends to introduce distortion, but, since it operates on such a small fraction of the signal, the effect of any distortion is negligible.

The characteristics of the recording processor and the replay processor should be exact opposites to obtain correct results. This is achieved by using the same processor for both purposes but just altering the connections, as shown in figure 11.59. On record the signals are processed by P and added to the original signal so that the high frequency signals below a certain value are increased in amplitude. On playback the signals are processed by the same processor P, but the signal output from the processor is now subtracted from the signal, so that the high frequency signals (below a certain value) are decreased to their original value. The processor may be built into

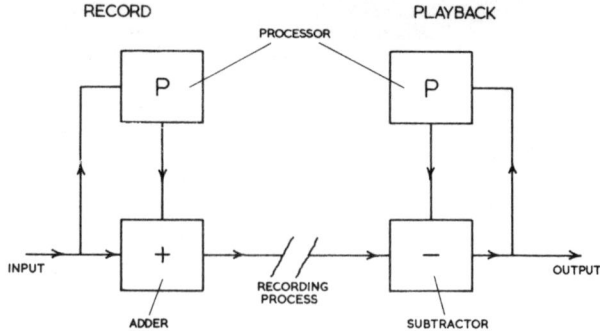

FIG. 11.59. DOLBY SYSTEM

the recorder (as it normally is) or a separate unit can be used. The system may be used with reel-to-reel or cassette machines, and an improvement of the signal-to-noise ratio of some 10 dB is obtainable. Some pre-recorded tapes are now available using Dolby recordings. They can be played back on a player not fitted with a Dolby processor (although the advantages of Dolby are lost) by turning down the treble tone control. The Dolby process does not, of course, reduce the noise in the original signal and must be applied to both recording and playback.

(b) Philips Dynamic Noise Limiter (DNL)

This equipment is used on replay only, but in fact it can be used to reduce the noise in any signal. It is based on the idea that, on soft passages of music, there are few higher harmonics produced, which is a property of musical instruments. Thus, during soft passages the frequency response can be reduced and will reduce the noise. The basic arrangement is shown in figure 11.60. The signal is split into two paths. The lower path is through an all-pass filter; hence the output is the same as the

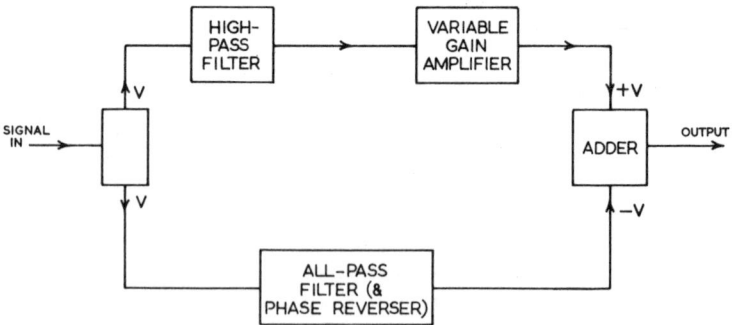

FIG. 11.60. PHILIPS DYNAMIC NOISE LIMITER (DNL)

input, except that the polarity of the signal is reversed. The upper portion consists of a high-pass filter and variable gain amplifier. This circuit is so designed that when the high frequency signals are of small magnitude, the gain is increased and these are added (in antiphase) in the adding circuit. Thus, on quiet passages, the high frequencies and noise cancel out, and so reduce the noise level. On loud passages it has little effect. Again it is important to note that the processing is only done on the high frequency signals. The improvement is claimed to be 10 dB at 6 kHz and 20 dB at 10 kHz.

CHAPTER 12

QUADRAPHONIC REPRODUCTION

THE use of stereo reproduction allows the listener to place the source of sound when it is at the front. However, it does not enable one to locate sounds that might occur at the back. In fact, there is no source of sound at the rear. The idea of quadraphonic reproduction is to enable the listener to locate sounds throughout the 360°. Before going on to quadraphonics we will first consider what is known as "ambio reproduction", or sometimes as "pseudo quadraphonics".

AMBIO REPRODUCTION (AMBIOSONICS OR SURROUND-SOUND)

One might first say, why do we wish to reproduce sounds at the rear of the listener. If one considers the reproduction of an orchestra in a hall, the listener in the hall hears sounds reflected from the back of it, which gives "presence" to the sound, *i.e.* it enables him to realise that he is listening in a hall. Even with stereo this is not reproduced in a normal room. There will, of course, be reflections from the back of the room, but, due to the size of the room, they are quite different from those heard in, say, a large hall.

In ambio reproduction two speakers are placed at the back corners of the room, as in figure 12.1, these speakers being fed with the difference signal, *i.e.* L–R (and R–L). The back speakers are fed with signals 180° out of phase with each other as shown in

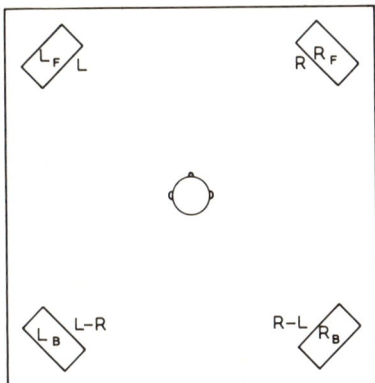

FIG. 12.1. AMBIO OR PSEUDO-QUADRAPHONIC REPRODUCTION

the figure. The system has problems but generally does add a pleasant effect, depending on the music being reproduced. One problem arises when, say, most of the signal is on the left channel. In these circumstances the L front speaker and left L–R speaker both have the same signal (since R = 0), and the stereo image moves to somewhere between the L and L–R speakers, the movement of the stereo image now being emphasized. Also, the signal-to-noise ratio of the difference signal is less than that of the L or R signal, but improvement can result if the high frequency response of the rear speakers is reduced (see figure 12.2). The advantage of this scheme is that it is simple and relatively cheap (the additional cost is two speakers, which need not be as good as the front speakers). The speakers are simply connected as shown in figure 12.2. Some stereo amplifiers have sockets so that the additional speakers can be used. It may be desirable to fit a preset resistor in the rear speaker circuit so that the relative volumes

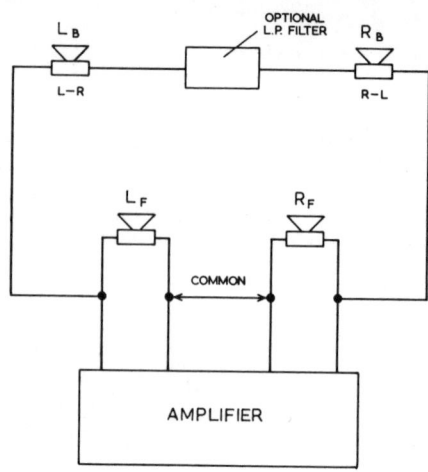

FIG. 12.2. CONNECTION OF LOUDSPEAKERS FOR AMBIO REPRODUCTION

of front and rear speakers can be adjusted as desired.

In Chapter 9 it was stated that two directional microphones placed at right angles were used for recording. It is important to point out that many records are not made in this way, particularly "pop" records. It is common practice to have 8 to 24 channel recorders with a number of microphones, and one may consider the extreme case of one microphone per instrument. The instruments may not be recorded at the same time and not even in the same studio. The final record is obtained by mixing these tracks as required, the volume by means of volume controls and the position by what is known as "pan pots". By varying the relative signals to L and R channels the position of the instrument can be moved as required. The final sound produced may, in fact, be one that is quite impossible to produce by being present at the studio. In a similar way it is possible to add sound from, say, the back of a hall, as a difference signal into the recording. When ambio reproduction is used then this will appear in the rear speakers.

QUADRAPHONY REPRODUCTION

The basic idea is to have four microphones, preferably with responses like those shown in figure 12.3, the microphones being at 90° to each other. These microphones are now connected to four loudspeakers at the corners of a square with the listener in the centre. One might now expect that any sound in the studio will be reproduced so that the listener can locate the sound over the whole 360°. Normally one considers the microphones and loudspeakers as all being at the same vertical level, but this is not essential and they may be arranged in various positions along the sides or corners of a cube. This is known as tetrahedral reproduction. Much work, both experimental and theoretical, has gone into quadraphonic reproduction and the effects on the listener, but we will assume that this basic idea works satisfactorily. Like stereo recordings, most quadraphonic recordings are made using multi-track recorders. It is much easier to move an instrument by using a "pan pot" than to move the player and then record the music again. Once satisfactory multitrack recordings have been made the players can leave and the final recording can be produced much later. It is very easy to try various combinations and effects until the required result is obtained.

One may ask why a listener wants a quadraphonic recording. It will, of course, add the sound reflections from the various parts of the hall or studio as mentioned earlier. It can, of course, be used for spectacular effects, such as car racing, but not all of us want to listen to cars racing round and round the room. It can, and often does,

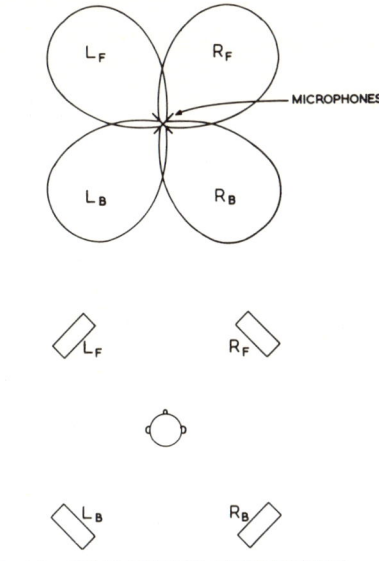

FIG. 12.3. QUADRAPHONIC REPRODUCTION

give the impression of one being in the centre of the band or orchestra, but again this is not what many people want. If one goes to a classical concert in a hall then, unless one is near the front, the stereo angle is small and probably much less than that normally used in stereo reproduction. The reflected sound from the sides and back will be added by using quadraphonic reproduction, the magnitude and importance of these will vary greatly with the hall and would, for example, be greatest in a cathedral. One may, particularly with "pop" music, wish to produce a pleasant sound or experience which is not what one would obtain as a member of the audience to the "pop" group. If this is the case then quadraphonic reproduction is fine.

The other disadvantages are expense and space. Four amplifiers and speakers are required, instead of two as in stereo, and one may also require a decoder of some type (see later). Also space is necessary to put the four speakers in suitable positions in the room. Stereo is often difficult and quadraphonic reproduction is much more difficult. The seating area for satisfactory reproduction may also be reduced compared with stereo.

METHOD OF RECORDING

We will assume that four signals are available at the studio and that four amplifiers and speakers are also available in the listening room. How are the signals transferred from the studio to the listener? There are three basic methods: by tape, by record, or by radio. In the case of tape then four channels are necessary and can be provided; at least one tape recorder (TEAC) has four independent channels using four tracks ($\frac{1}{4}$-track recording). This seems the obvious way, and presents no real difficulties apart from cost and the fact that pre-recorded tapes (other than cassettes and cartridges) have never been popular. The tape would be relatively expensive (at least twice the cost of a stereo tape because twice the amount of tape is used). A four-channel recorder is expensive (simply because there is four of everything) and if such tapes were sold they would only be suitable for those with four-channel recorders, which means that listeners could not use their present equipment (it might be possible to play stereo off a four-channel tape depending on how it was recorded). Hence, this idea does not seem popular, although technically it is the correct answer, and, of course, used for the master in other methods. It is known as a discrete system, since the four channels are kept separate throughout. It is sometimes called a:

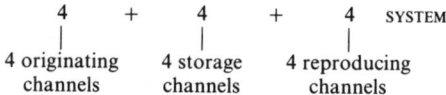

The ambio or pseudo system is sometimes called a:

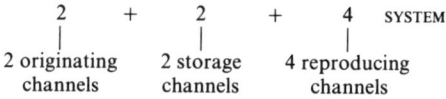

It should be mentioned that there are some cartridges available for quadraphonic reproduction intended for cars. (One would doubt the need for quadraphonic reproduction in a car). In this case four of the available tracks are used for one programme and the other four for the other programme. For the arrangement see figure 12.4. One firm has produced cassettes and a recorder with four tracks in place of the two for stereo (*i.e.* a total of eight instead of four). The tracks are obviously extremely

PROGRAMME 1	PROGRAMME 2	TRACK
L_F		1
	L_F	2
L_R		3
	L_R	4
R_F		5
	R_F	6
R_R		7
	R_R	8

FIG. 12.4. TRACKS OF TAPE OF QUADRAPHONIC CARTRIDGE

narrow, and the author feels that this is making a difficult problem for recorder manufacturers. Why not use the present four tracks and play the cassette in one direction only? The author has not heard of such a system being suggested commercially. It would have the difficulty that the cassette would have to be re-wound each time it was used.

The most common recording device is the record, and this is where the problem arises. Many systems have been devised to get quadraphonic recordings on a record and some of these will be described. Obviously there are many large recording companies and others with large financial interests in the development of these systems, and it is hoped that some common system will eventually be evolved.

CD-4 SYSTEM

This is considered first as it is a $4+4+4$ system. The problem is, how does one record 4 channels on a single groove of a gramophone record? The system was developed by JVC and RCA. Only the basic principles of the system will be described. To carry the additional information a carrier is used, being frequency modulated. The idea is shown in figure 12.5. Consider L_F as left front channel, R_F as right front,

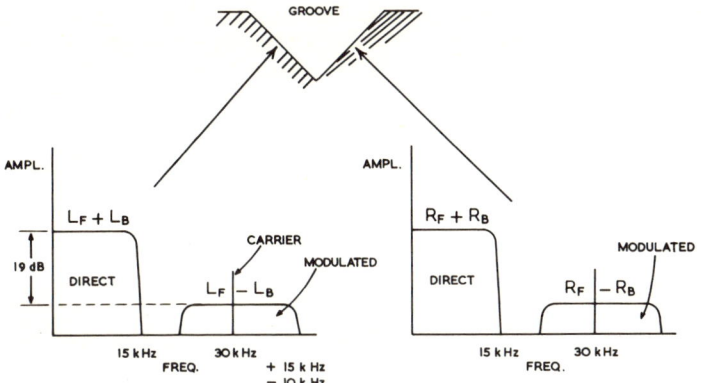

FIG. 12.5. CD-4 QUADRAPHONIC SYSTEM

L_B as left back and R_B as right back channel. On one wall of the groove is placed:

$L_F + L_B$ up to a frequency of 15 kHz
$L_F - L_B$ modulated on to carrier of 30 kHz.

On the other wall of the groove is placed:

$R_F + R_B$ up to a frequency of 15 kHz
$R_F - R_B$ modulated on to carrier of 30 kHz.

Actually simple frequency modulation is not used. Up to 800 Hz the carrier is frequency modulated, from 800 Hz to 6 kHz it is phase modulated, and above 6 kHz it is frequency modulated. The recording equipment is extremely complex as noise reduction circuits are used on the difference signals, and devices are used to prevent modulation of the carrier by the main signal. Perhaps the complexities of recording are not important provided the reproducing equipment is simple. The reproducing equipment is certainly simpler than that for recording, but still complex. First, one must have a pick-up that will operate up to, say, 45 kHz, which, at first, sounds impossible. However, pick-up cartridges are now available, some using the Shibata stylus, which has particular advantages at this high recording frequency. The signals from the pick-up must be sorted out in a matrix and the modulated carriers demodulated to recover the various signals. An integrated circuit is being developed for this purpose. Since the normal modulation of the side walls are L and R (both back and front) then a reasonable stereo reproduction will be obtained from a stereo reproducer.

A variation of this idea has been introduced by Nippon Columbia called the UD-4 system. This system uses the same basic ideas but the signals carried by the four channels are different, a matrix being used before recording. It is claimed that this system will give normal mono and stereo reproduction, 2 channel matrix system (see later) and 4 channel reproduction. This means that the equipment will play records using any one of the matrix systems to be described.

MATRIX SYSTEMS

These are systems which only use 2 storage channels, or are

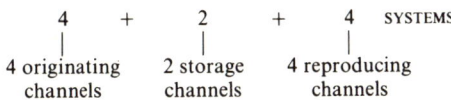

The idea of these systems is to code or matrix the 4 channels into 2 channels for storage (*i.e.* the record) and convert it back again into 4 channels. Obviously, some information is lost in the original matrixing to 2 channels and cannot be recovered. There are several systems, the most common being the SQ system (Stereophonic-Quadraphonic) developed by CBS. The way the four signals are matrixed is given in figure 12.6.

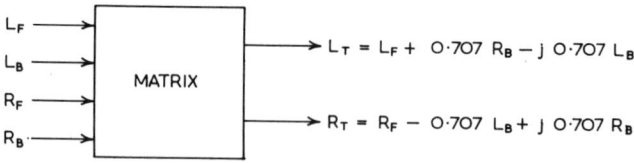

$$L_T = L_F + 0.707\ R_B - j\ 0.707\ L_B$$
$$R_T = R_F - 0.707\ L_B + j\ 0.707\ R_B$$

FIG. 12.6. SQ MATRIX SYSTEM OF QUADRAPHONICS—CODING

L_T means the total signal to modulate one wall and R_T that modulating the other wall of the groove of the record. The j means a phase shift of $+90°$. The corresponding phasor diagrams are given in figure 12.7. It will be noted that the L_F signal only appears in the L_T signal, and similarly the R_F signal only appears in the R_T signal.

The decoder for this recording system is given in figure 12.8.

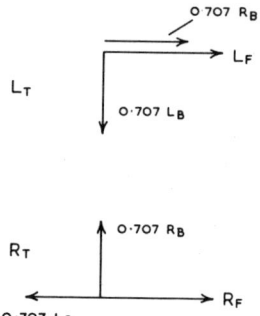

FIG. 12.7. SQ SYSTEM—PHASOR DIAGRAM OF CODING

The figures with primes (') mean the recovered channels. Now L'_F is, of course, the same as $L_T = L_F + 0.707\ R_B - j\ 0.707\ L_B$. Similarly R'_F is the same as $R_T = R_F - 0.707\ L_B + j\ 0.707\ R_B$. Consider now L'_B. First the L_T signal is shifted by 90° lagging (i.e. multiplied by $-j$ and $j^2 = -1$),

so that L_T becomes $-j\ L_F - j\ 0.707\ R_B - 0.707\ L_B$

Adding R_T R_F $+j\ 0.707\ R_B - 0.707\ L_B$

results in $R_F - j\ L_F$ $-1.414\ L_B$

Multiplying by -0.707 this becomes $-0.707R_F + j\ 0.707L_F + L_B$ the same as that shown in figure 12.8 (but in a different order). A similar calculation gives R'_B. The resulting phasor diagrams are shown in figure 12.9. Obviously these are far from the original L_F, L_B, R_F and R_B signals, and there is considerable crosstalk. If there is only an L_F original signal (or R_F signal) then the result is correct because the L'_F and R'_F become L_F and R_F (the other signals being zero). If there are also some L_B and R_B signals these will appear in both front channels, i.e. there is $-j\ 0.707L_B$ in L'_F and $-0.707L_B$ in the R'_F channels. The two signals are in quadrature and in practice (because they are in quadrature), the signals appear in the distance and in no clearly defined position. The worst crosstalk results from a source in the centre front where $L_F = R_F$. Under these conditions the front signals are correct (since R_B and L_B

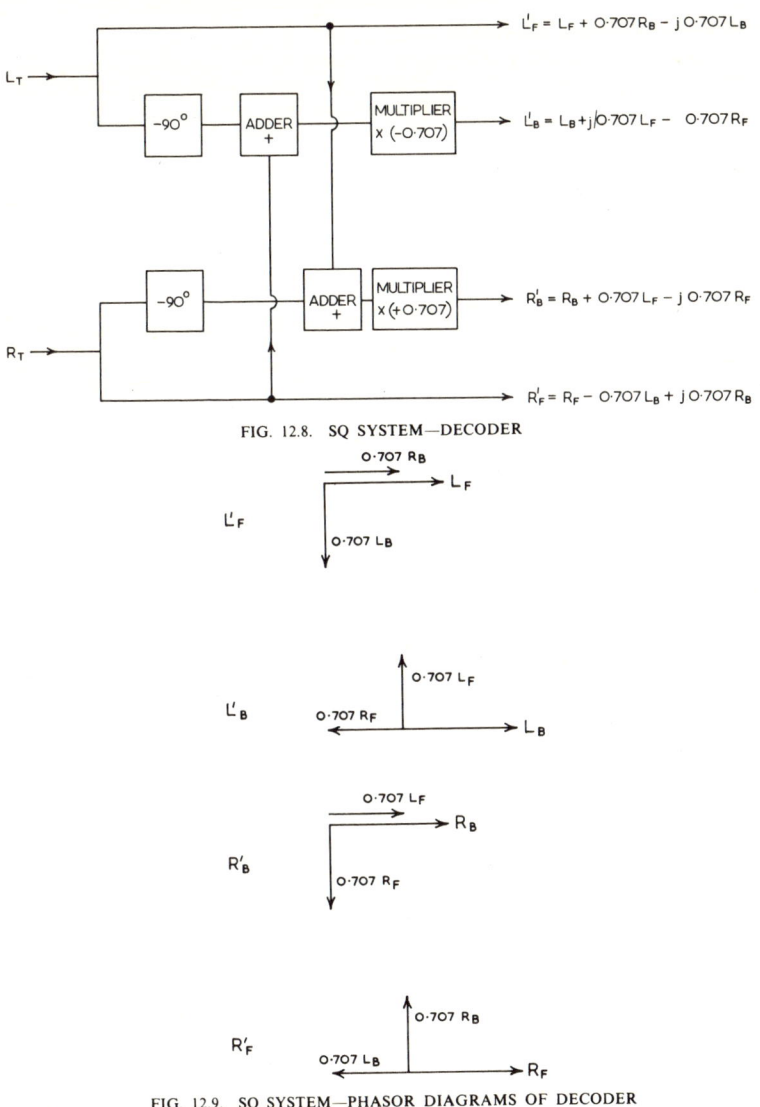

FIG. 12.8. SQ SYSTEM—DECODER

FIG. 12.9. SQ SYSTEM—PHASOR DIAGRAMS OF DECODER

are zero). However, the rear signals become:

$L'_B = j\,0.707\,L_F - 0.707\,L_F$ (substituting L_F for R_F)
$R'_B = 0.707\,L_F - j\,0.707\,L_F$ (substituting L_F for R_F)

The phasor diagrams for L'_B and R'_B are shown in figure 12.10. Thus L'_B and R'_B are numerically equal and in antiphase, and also of equal amplitude to the signals fed to the front speakers. Although they are in antiphase they do upset the central front positional image. This can be reduced by what is called "blending" between the two front outputs and the two rear outputs. Commonly, 10% blending or mixing of the front outputs together with 40% blending of the rear outputs is used. Hence the

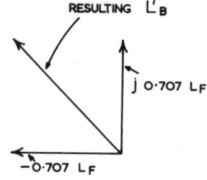

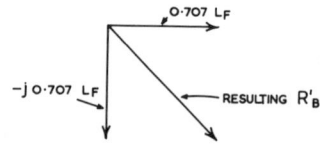

FIG. 12.10.   SQ SYSTEM WHEN $L_F = R_F$

new signals (double primed) fed to the speakers become:

$$R''_F = 0.9\ R'_F + 0.1\ L'_F$$
$$L''_F = 0.9\ L'_F + 0.1\ R'_F$$
$$R''_B = 0.6\ R'_B + 0.4\ L'_B$$
$$L''_B = 0.6\ L'_B + 0.4\ R'_B$$

All this decoding process can be done in a single integrated circuit (MC 1312).

To reduce some of the crosstalk effects a much more complex arrangement can be used. This detects errors by using a logic circuit (i.c.) which controls the gain of four amplifiers (i.c.) so as to reduce the error. As this is so complex no further details can be given.

One advantage of this system is that it is claimed to be compatible with mono and stereo, *i.e.* the disc can be played on a mono or stereo reproducer and produce satisfactory mono and stereo results.

It can be shown that this coding results in an additional helical modulation of the record groove.

Another system QS–RM is also in use developed by Sansui. Although the equations and the crosstalk problems are different, the general idea remains the same.

As these systems use the same bandwidth as mono they can be used for 2 channel tape recordings.

QUADRAPHONIC BROADCASTING

In a similar way to the development of stereo broadcasting, experiments are being carried out on systems for quadraphonic broadcasting. There are a number of experimental systems, but as these are only experimental no details will be given. Some use another subcarrier outside the present sound band, *i.e.* outside the band 0–53 kHz. One channel can be transmitted in the 23–53 kHz band by the use of quadrature modulation. The BBC have done some experimental quadraphonic broadcasts (outside normal broadcast hours) using two stereo channels. Although interesting this is not a practical solution owing to the use of twice the bandwidth.

The NQRC (National Quadraphonic Radio Committee) is at present looking into the problem of quadraphonic broadcasting and should report to the FCC (America) in 1975.

PRESENT POSITION

The present position of quadraphonic reproduction is very confused, and it is difficult to imagine large sales until a common international system is agreed on.

Obviously, large sums of money have been spent by various major companies in developing their own systems, and it is hard to see how a satisfactory solution can be arrived at. The matrix systems are far from perfect and do not give correct quadraphonic reproduction. One will have to wait and see if the public consider quadraphonic worth the considerable increase in cost. Compared with stereo, one requires two more amplifiers, two more speakers, and either a 4-track tape recorder or complex decoding system. If only synthetic sounds are required (rather than correct quadraphonic reproduction) then the matrix systems are probably satisfactory.

APPENDICES

APPENDIX 1

Power input, power output and transistor dissipation in class-B stage.

Let d.c. supply voltage be V_d (for one transistor)
Let peak a.c. output voltage be V_p
Let peak a.c. output current be I_p

Considering figure 3.8 the instantaneous collector voltage $v = V_d - V_p \sin \theta$

The instantaneous collector current $i = I_p \sin \theta$

Power dissipated per transistor (transistor only conducts for a half-cycle)

$$= \frac{1}{2\pi} \int_0^\pi v i \, d\theta$$

$$= \frac{1}{2\pi} \int_0^\pi (V_d - V_p \sin \theta)(I_p \sin \theta) \, d\theta$$

$$= \frac{1}{2\pi} \int_0^\pi V_d I_p \sin \theta - \frac{V_p I_p}{2}(1 - \cos 2\theta) \, d\theta$$

$$= \frac{1}{2\pi} \left[-V_d I_p \cos \theta - \frac{V_p I_p}{2}\theta + \frac{V_p I_p}{2} \times \frac{\sin 2\theta}{2} \right]_0^\pi$$

$$= \frac{V_d I_p}{\pi} - \frac{V_p I_p}{4}$$

Power dissipation in both transistors $= \dfrac{2V_d I_p}{\pi} - \dfrac{V_p I_p}{2}$ \hfill **(A.1)**

$\dfrac{2V_d I_p}{\pi}$ is the power input; and

$\dfrac{V_p I_p}{2}$ is the power output

Substituting the fact that $I_p = \dfrac{V_p}{R}$

Power dissipation in both transistors:

$$= \frac{2V_d V_p}{\pi R} - \frac{V_p^{\,2}}{2R}$$ \hfill **(A.2)**

$\dfrac{2V_d V_p}{\pi R}$ is the power input and $\dfrac{V_p^{\,2}}{2R}$ is the power output.

This power dissipation is a maximum when differentiated with respect to V_p and equated to 0.

i.e. $\dfrac{2V_d}{\pi R} - \dfrac{2V_p}{2R} = 0$

or $V_p = \dfrac{2}{\pi} V_d = \underline{0.636 V_d}$

Thus the dissipation is not a maximum at maximum power output but when $V_p = 0.636 V_d$.

197

APPENDIX 2

Relationship between voltage (and power) ratio and decibels.

In figure A.1 the relationship is plotted between voltage ratio and decibels. When the ratio is greater than unity the lower horizontal scale and the left-hand vertical scale should be used, the answer being positive decibels. When the ratio is less than unity the top horizontal scale should be used, together with the right-hand vertical scale, the answer now being in negative decibels.

The same graph can be used for power ratio but in this case the reading on the dB scale must be halved. For example, a voltage ratio of 10 results in 20 dB but a power ratio of 10 results in 10 dB.

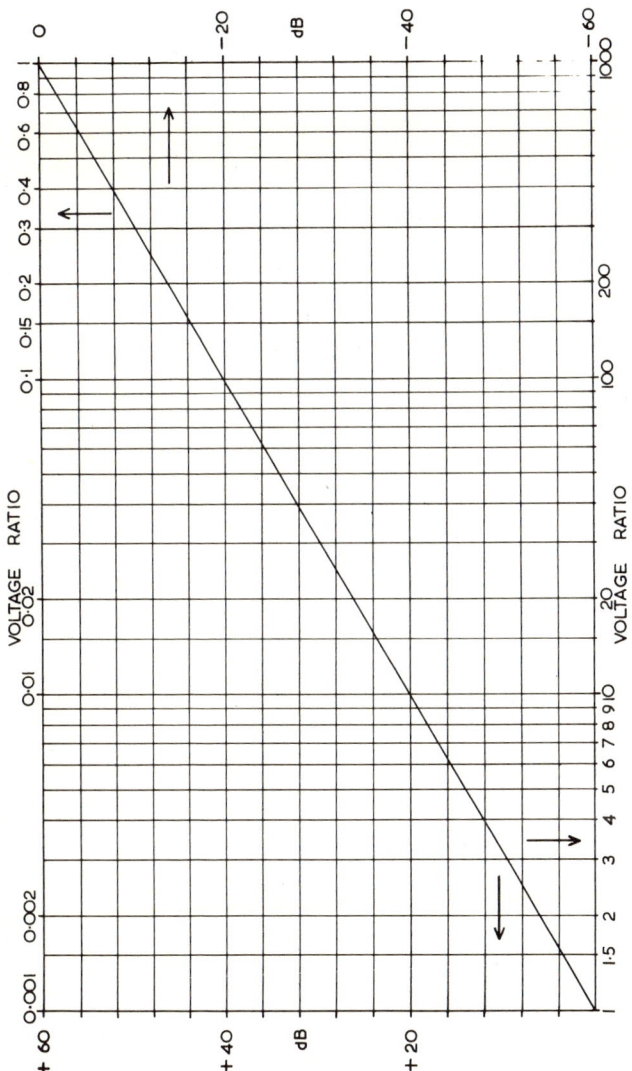

FIG. A.1   RELATIONSHIP BETWEEN VOLTAGE RATIO AND dB.

APPENDIX 3

Reactance of capacitors at various frequencies (in ohms).

CAPACITANCE VALUE

FREQUENCY (Hz)	10 pF	100 pF	1000 pF	10,000 pF	100,000 pF	1 μF	10 μF
10	—	160 M	16 M	1·6 M	160 K	16 K	1·6 K
100	160 M	16 M	1·6 M	160 K	16 K	1·6 K	160
1000	16 M	1·6 M	160 K	16 K	1·6 K	160	16
10,000	1·6 M	160 K	16 K	1·6 K	160	16	1·6
100,000	160 K	16 K	1·6 K	160	16	1·6	0·16
1 × 10^6	16 K	1·6 K	160	16	1·6	0·16	—
10 × 10^6	1·6 K	160	16	1·6	0·16	—	—
100 × 10^6	160	16	1·6	0·16	—	—	—

APPENDIX 4

Reactance of inductors of various frequencies (in ohms).

(Pure inductors are assumed)

INDUCTANCE VALUE

FREQUENCY (Hz)	10 μH	100 μH	1 mH	10 mH	100 mH	1 H	10 H
10	—	—	—	0·63	6·3	63	630
100	—	—	0·63	6·3	63	630	6·3 K
1000	—	0·63	6·3	63	630	6·3 K	63 K
10,000	0·63	6·3	63	630	6·3 K	63 K	630 K
100,000	6·3	63	630	6·3 K	63 K	630 K	—
1×10^6	63	630	6·3 K	63 K	630 K	—	—
10×10^6	630	6·3 K	63 K	630 K	—	—	—
100×10^6	6·3 K	63 K	630 K	—	—	—	—

INDEX